SpringerBriefs in Physics

More information about this series at http://www.springer.com/series/8902

Igor Minin · Oleg Minin

Diffractive Optics and Nanophotonics

Resolution Below the Diffraction Limit

Springer

Igor Minin
Tomsk State University
Novosibirsk
Russia

Oleg Minin
Tomsk State University
Novosibirsk
Russia

ISSN 2191-5423 ISSN 2191-5431 (electronic)
SpringerBriefs in Physics
ISBN 978-3-319-24251-4 ISBN 978-3-319-24253-8 (eBook)
DOI 10.1007/978-3-319-24253-8

Library of Congress Control Number: 2015950920

Springer Cham Heidelberg New York Dordrecht London

Printed on acid-free paper

Springer International Publishing AG Switzerland is part of Springer Science+Business Media
(www.springer.com)

Dedicated to the father, friend and colleague, all in one, and equally to our mother. Without their help and support, this book would never have been written.

Foreword I

The major limitation of contemporary imaging systems is based on the diffraction limit of light. The physical meaning of this limit is that the light cannot be squeezed into the dimensions smaller than its wavelength (λ). More specific Abbe, Rayleigh, Sparrow, and Houston resolution criteria show that the limit of lateral resolution is close to a half of the wavelength. The resolution can be further improved by immersing an object in a medium with refractive index (n). This allows achieving the resolution at the $\lambda/2n$ level, but this is a limit of the far-field optics indicating that further increase of the resolution requires different physical principles.

This limitation is the most important bottleneck problem of imaging in nanoscale science and technology, which operates with the objects smaller than 100 nm in size. It means that the visible light with $400 < \lambda < 700$ nm cannot be focused below 100 nm or used for imaging of objects with better than 100 nm resolution. However, the most important objects of nanoscale and life sciences are smaller than 100 nm in size. These include internal cellular structures, DNA and RNA, viruses, proteins, quantum dots and wires, fullerenes, carbon nanotubes, fluorophores, dye molecules, etc.

Due to its fundamental nature, the diffraction limit forms an unavoidable barrier for any far-field imaging system. Increasing the resolution beyond the diffraction limit usually requires either detecting of the evanescent waves which exist only in near-field (in nanoscale proximity to the object's surface), or using strong optical nonlinear effects, or obtaining additional information about objects by other methods. Description of these techniques goes beyond the scope of this book, and it constitutes very active area of modern photonics. One of the examples of importance of this area is represented by awarding of 2014 Nobel Prize in Chemistry to Eric Betzig, Stefan Hell and William E. Moerner for their pioneering work in "super-resolution" fluorescence microscopy.

This book puts forward a more confined, but very important task. It systematically studies the optical properties of *mesoscale* lenses with dimensions comparable to the wavelength of light: Fresnel zone plates (FZP), photonic crystal structures and dielectric objects with various shapes. All these lenses are designed to focus light in the near-field vicinity of their surfaces and that, in principle, opens a possibility of overcoming the diffraction limit. It should be noted, however, that the conventional analytical methods are not well suited for describing an extremely complicated interplay of various optical effects in mesoscale objects. This interplay includes the coexistence of near- and far-fields, interference of zero-order transmitted and scattered beams, and significant role played by the polarization and directional properties of the incident beams. In addition, the optical near-fields can be resonantly enhanced in such mesoscale objects. This dictates a computational approach of this book based on numerical solution of the Maxwell's equations.

This book is mainly built around the microwave and terahertz applications; however its approach can be also viewed as a scaled model at other frequencies including the optical range. The results indicate that in many cases the subdiffraction-limited beam waists can be realized close to the surfaces of such objects. Although the advancement over conventional diffraction limit is not dramatic (typically limited by $\lambda/3$ in air), this is an important step for developing nanoscience and nanotechnology applications. As an example, such mesoscale structures can be easily fabricated and, in the case of liquid or solid immersion, they can provide nanoscopy with the resolution better than 100 nm using visible light. Such mesoscale lenses can be integrated, for example, into existing semiconductor heterostructure platforms. In addition, these structures can be used as tips of new generations of scanning near field optical microscopes or focusing microprobes for ultraprecise laser surgery.

In the first three chapters, the authors analyze different types of mesoscale lenses. In the fourth chapter, they extend some of their designs to the structures supporting surface plasmon-polaritons (SPPs) with much shorter wavelength compared to that in air. This chapter shows a large potential of such nanoplasmonic structures for increasing resolution. It is likely that the concept of the in-plane SPP curvilinear FZP-like lens will have a significant impact in science and technology.

Throughout the whole book, the reader can recognize that I. Minin and O. Minin are very experienced researchers not only with respect to the scientific quality of their results but especially with regard to the didactic approach of this textbook. The layout of the book is functional and sober, as one expects from a physics monograph. This book is a creative collection of rigorous scientific research and very practical examples. It provides several new insights in the field of optical elements with super-resolution focus and also contains many interesting results that are worth applying in practice, while it is also a source of new and intriguing questions for further research.

Prof. Vasily Astratov
Department of Physics and Optical Science
at the University of North Carolina-Charlotte, USA

Foreword II

Since the discovery of electromagnetic waves about 150 years ago by J.C. Maxwell, millimeter-waves and terahertz waves, which are located between microwaves and infrared light waves, have been untapped in our life until this century. The research and development of these electromagnetic waves have lately being focused on real-world applications, which include material spectroscopy, nondestructive inspection of objects, security imaging as well as communications. Such exciting waves have also attracted me for over 30 years. This book is dedicated to overcoming classical diffraction-limited issues we are facing in controlling waves. The principles and concepts described in the book are expected to enhance performances in sensing, measurement, and communications applications in the millimeter wave and terahertz regions as well as infrared light waves.

Professors Igor Minin and Oleg Minin are eminent researchers in the topical areas of diffractive optics, antenna theory, millimeter wave and terahertz photonics, plasmonics, and so on, and they have actively published several books related to these topics. This new book covers one of their most recent challenges on "subwavelength focusing," which could break through the well-known "diffraction limit" discovered by E.R. Abbe a few years after the Maxwell's feat. In this book, they introduce novel ideas of nanophotonics, such as 2D/3D diffractive optical element, photonic crystal lenses, photonic jets, and surface plasmon diffractive optics, in order to control millimeter waves and terahertz waves. For example, "terajets" which consist of 3D dielectric cuboids, and "teraknife" with 2D dielectric rods provide practical tools for the subwavelength focusing at terahertz frequencies.

The publication of this book is very timely and of great benefit for those who are involved in the field of both radio waves and light waves, or microwaves and photonics. It is anticipated that the book will stimulate innovative ideas to advance millimeter-wave and terahertz technologies not only in devices but also in integrated systems.

Prof. Tadao Nagatsuma
Osaka University, FIEEE, FIEICE, FEMA

Acknowledgments

We are greatly indebted to Dr. Aldo Petosa and Dr. N. Gagnon (CRC, Canada) for their contribution to simulation of FZP with short focus and also to Dr. Miguel Beruete and Victor Pacheco (Antennas Group-TERALAB, Universidad Pública de Navarra, Spain) for help in simulations and discussions in our joint works on cuboid particles. We would like to thank Borislav Vasić (Institute of Physics Belgrade, Serbia) for help in simulations of metacuboid. We are infinitely grateful to Profs. Tadao Nagatsuma (Osaka University, Japan) and Vasily Astratov (Department of Physics and Optical Science at the University of North Carolina-Charlotte) for their willingness to write the Forewords to this book and also to Tom Spicer at Springer, for constant attention and help in our work with the book.

Contents

Chapter 1
Introduction

Abstract The criterions of resolution based on classical theory are discussed briefly. The main definitions of focusing area are described.

Keywords Resolution power · Abbe criterion · Rayleigh criterion · Heisenberg's principle

The progress in millimeter wave/THz/optical technology and photonics industry became difficult due to a fundamental limit of radiation known as the diffraction limit.

In 1873, the German physicist Ernst Karl Abbe (Fig. 1.1) discovered a fundamental 'diffraction limit' (Latin *diffractus* means fractured, broken, and obstacle avoidance by waves) in optics [1, 2]. From this law it was followed that radiation cannot be focused into a point. The size of the minimal focus spot by the radiation intensity decay halftime equals half the wavelength in the medium under consideration.

For diffraction-limited optical and quasi-optical systems, which are subject to the paraxial approximation, the Rayleigh criterion gives the spatial resolution, Δ, for a circular lens of diameter D and focal length F as [3]:

$$\Delta = \Delta_0 \lambda \frac{F}{D} \tag{1.1}$$

where $\Delta_o = 1.22$ is called a resolution coefficient. The resolution (1) was found for a circular aperture illuminated normally by a plane wave.

In terms of resolution, the radius of the diffraction Airy disk in the lateral (x,y) image plane (in the scalar approximation the full-width at half-maximum (FWHW) spot size) is defined by the following simple formula [4, 5]:

$$\mathrm{R_e} \sim \lambda/2\mathrm{NA} \tag{1.2}$$

As a result of this relationship, the size of the spot created by a point source decreases with decreasing wavelength and increasing numerical aperture. For air $n = 1$ and high objective aperture angle ($\sin(\theta) = 1$) the spatial resolution limit of objective lens are limited by diffraction and to be approximately half a wavelength.

I. Minin and O. Minin, *Diffractive Optics and Nanophotonics*,
SpringerBriefs in Physics, DOI 10.1007/978-3-319-24253-8_1

Fig. 1.1 Ernst Karl Abbe and a fundamental law of 'diffraction limit'

The FWHM of the field intensity distribution in Airy disk for ideal optical system (without aberrations) is given by FWHM = 0.51λ/NA. But this equation for the amplitude distribution in the focal plane of a circular lens, is valid only for the large focal ratios F/D common in optical or quasi-optical systems.

Lord Rayleigh [3] gave us the well-known definition of spatial resolution, based on the separation of Airy discs: *two objects can then be resolved if the point of maximum intensity of one object falls onto the first minimum of the intensity distribution of the other object*. To a small extent, the Rayleigh criterion links resolution to signal-to-noise: the proposed separation of the intensity maxima corresponds to a drop of intensity from 100 to 73.5 % between the objects (for circular objectives). From (1), it is seen that the spatial resolution decreases linearly with *F/D*. However, the paraxial approximation is generally agreed to be accurate when the value of $F/D \geq 0.5$ [6], for which the spatial resolution is limited by

$$\Delta = 0.61\lambda/\text{NA}. \tag{1.3}$$

The best possible NA is NA = n which, for optical glasses or microwave dielectric material, is n $\approx$ 1.5 and hence $\Delta \approx \lambda/3$.

It is important to note that Abbe's and Rayleigh's criteria make no use of any information that is available on the properties of the two emitters. Furthermore, these criteria assume that the radiation-matter interaction is linear.

It could be also noted that the elongated geometry of the point-spread function[1] along the optical axis arises from the nature of the non-symmetrical wavefront that

[1]The 3D intensity distribution of the actual image in optics is called the Point Spread Function of a lens.

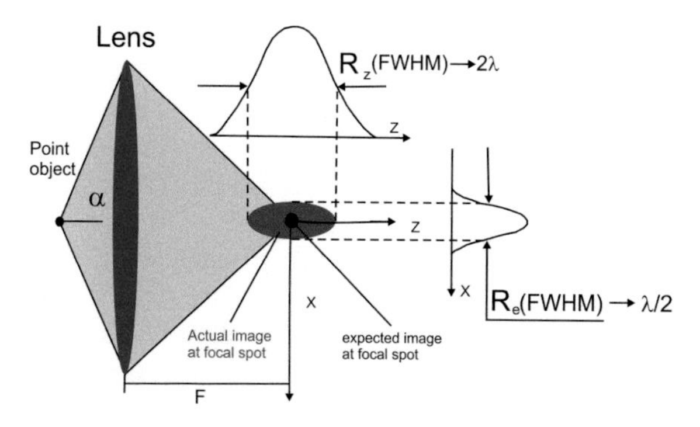

Fig. 1.2 Principle of image formation. This schematic depicts the intensity distribution from a object smaller than the wavelength of radiation. Note due to diffraction the actual image is much largest as well as elongated compared to the expected magnified image. The cross section profiles of this intensity distribution along X and Z axes are also shown. For air $n = 1$ and high objective aperture angle NA = 1 the limit of Abbe axial resolution is equal to 2 of wavelength and the spatial resolution limit is approximately half a wavelength. The figure is not to scale

emerges from the lens. Axial resolution R_z according to Abbe's theory in lens is even worse than lateral resolution and is given by equation:

$$R_z = 2\lambda / NA^2 \tag{1.4}$$

For air $n = 1$ and high objective aperture angle $sin(\theta) = 1$ the limit of Abbe axial resolution is equal to 2 of wavelength (Fig. 1.2). Some discussion about numerical criteria of point image evaluation and image quality in millimeter wave/THz may be found at [7].

It could be also noted that the diffraction limit is often associated with Heisenberg's uncertainty principle [8]. We may to recognize the maximum transverse wave number as $Max(k_{II}) = \pm k^* \sin\theta_{max}$ which defines the bandwidth of spatial frequencies as $\Delta k = Max(k_{II}) = 4\pi NA/\lambda$. It is followed that $\Delta \times \Delta k = 0.614\pi$. In agreement with the uncertainty principle this product is larger than 1/2. It could be important to note that from Heisenberg's uncertainty principle it is also followed that the maximum permissible size of the optics is always greater than the linear size $\Delta x \approx \lambda/4\pi \approx \lambda/12$, which is approximately an order of magnitude smaller than the wavelength, and the corresponding diffraction limit.[2] It is followed that the maximum resolution corresponds to the minimum contribution of photons from the central part of the lens. And you must use a relative diameter ~ 10 of a lens and provide good monochromatic radiation. It should be noted that the presented

[2]The Rayleigh criterion is satisfied when the distance between the images of two closely spaced point sources is approximately equal to the width of the point-spread function. In contrast, the Sparrow resolution limit is defined as the distance between two point sources where the images no longer have a dip in brightness between the central peaks, but rather exhibit constant brightness across the region between the peaks and approximately equal to two-thirds (0.47 in contrast to 0.61) of the Rayleigh resolution limit.

conclusions are also valid for the electron and ion microscopy, where the wavelength must take de Broglie wavelength of an electron or ion.

The diffraction limit could be overcome using the novel technology of diffractive optics, nano-optics[3] or nanophotonics in which the size of the electromagnetic field is decreased down to the nanoscale and is used as a carrier for signal transmission, processing, and fabrication [9].

In opto- and microelectronics, as well as in nanophotonics use is made of sophisticated optical devices with the dimensions of the order of the wavelength of incident radiation, whose work is described by the non-trivial physical effects, such as, for example, the scattering and diffraction on aperiodic structures (like diffractive optical elements (DOE)). The effect of these elements can not be predicted on the basis of geometrical optics or scalar diffraction theory, and it is essential to study the propagation of electromagnetic waves through them using the Maxwell equations.

It could be noted that for the problems of modelling the radiation diffraction on the elements with dimensions on the order of the wavelength the radiation must be regarded as electromagnetic radiation, which allows us to transfer many of the developed methods of electromagnetic simulation of microwave and radio waves [7, 10] to the area of optical modelling.

To overcome the problem of the diffraction limit, different solutions have been reported in the past, for example, involving metamaterials [11–13], solid immersion lenses [14], diffractive optics [15, 16] and microspherical particles [17–19]. But what is the limit that can be "overcome"?

Below we will briefly consider several examples of overcoming the diffraction limit using: flat and 3D diffractive optical elements, photonic crystal lenses, photonic jets, and surface plasmon diffractive optics. The structure here discussed can be used at microwaves/THz and also as a scaled model (because of all dimensions are given in the wavelength units) at optical frequencies. Such nano-optical microlenses can be integrated, for example, into existing semiconductor heterostructure platforms for next-generation optoelectronic applications.

In the conclusion, the main directions of development of diffractive optics and nanophotonics according to the authors' point of view are discussed.

In the book we briefly described different types of focusing diffractive elements with subwavelength resolution. We suggest a mesoscale (the size of the particles can be of the order of λ or larger) 3D dielectric structure whose properties are determined not only by its inner geometry and materials property but also by its entire 3D shape. The design of mesoscale optical materials and structures is the focus of much current multidisciplinary research. We call this structure a *mesoshape*. We evaluate the potential of mesoshape to control both the size and 3D location of the field enhancement with subwavelength resolution.

[3]Although terminology is not strict one can refer to *nanooptics* addressing more fundamental aspects and *nanophotonics* addressing more applied aspects, respectively (so nanooptics has emerged from the wider area of nanoscience).

This work not only reveals the explicit physical role of any given types of diffractive optics and nanophotonics, but to our point also provides an alternative design roadmap of superresolution imaging.

References

1. Abbe, E. (1873). Beiträge zur Theorie des Mikroskops und der mikroskopischen Wahrnehmung. *M. Schultze's Archiv für mikroskopische Anatomie, 9*, 413–468.
2. Abbe, E. (1880). Ueber die Grenzen der geometrischen Optik. *Jenaische Zeitschrift für Naturwissenschaft. Sitzungsberichte, 14*, 71–109.
3. Rayleigh, L. (1896). On the theory of optical images, with special reference to the microscope. *Philosophical Magazine 54*, 167.
4. Airy, G. B. (1835). On the diffraction of an object-glass with circular aperture. *Transactions of the Cambridge Philosophical Society, 5*, 283–291 (1835).
5. Airy, G. B. (1841). On the diffraction of an annular aperture. *Philosophical Magazine Third Series, 18*(114), 1–10.
6. Goldsmith, P. F. (1998). *Quasioptical Systems*. New York: IEEE Press.
7. Minin, I. V., Minin. O. V. (2008). *Basic principles of Fresnel antenna arrays*. Lecture Notes Electrical Engineering (Vol. 19). Berlin: Springer.
8. Novotny, L. (2007). The history of near-field optics. In E. Wolf (Ed.), *Progress in optics* (vol. 50, chapter 5, pp. 137–184). Amsterdam: Elsevier.
9. Ohtsu, M. (ed.). (2013). *Handbook of nano-optics and nanophotonics* (1071 p). Berlin: Springer.
10. Minin, O. V., & Minin, I. V. (2004). *Diffractional optics of millimeter waves*. Boston: IOP Publisher.
11. Pendry, J. B. (2000). Negative refraction makes a perfect lens. *Physical Review Letters, 85*, 3966–3969.
12. Fang, N., Lee, H., Sun, C., & Zhang, X. (2005). Sub-diffraction-limited optical imaging with a silver superlens. *Science, 308*, 534.
13. Liu, Z., Lee, H., Xiong, Y., Sun, C., & Zhang, X. (2007). Optical hyperlens magnifying sub-diffraction-limited objects. *Science, 315*, 1686.
14. Mansfield, S. M., & Kino, G. S. (1990). Solid immersion microscope. *Applied Physics Letters, 57*, 2615.
15. Minin, I. V., Minin, O. V., Gagnon, N., Petosa, A. (2006). FDTD Analysis of a Flat Diffractive Optics with Sub-Reyleigh Limit Resolution in MM/THz Waveband. In *Digest of the Joint 31st International Conference on Infrared and Millimeter Waves and 14th International Conference on Terahertz Electronics* (p. 170). Shanghai, China, September 18–22, 2006.
16. Minin, I. V., & Minin, O. V. (2014). 3D diffractive lenses to overcome the 3D Abbe subwavelength diffraction limit. *Chinese Optics Letters, 12*, 060014.
17. Heifetz, A., Kong, S.-C., Sahakian, A. V., Taflove, A., & Backman, V. (2009). Photonic nanojets. *Journal of Computational and Theoretical Nanoscience, 6*, 1979.
18. Chen, Z., Taflove, A., & Backman, V. (2004). Photonic nanojet enhancement of backscattering of light by nanoparticles: a potential novel visible-light ultramicroscopy technique. *Optics Express, 12*(7), 1214–1220.
19. Pacheco-Pena, V., Beruete, M., Minin, I. V., & Minin, O. V. (2014). Terajets produced by 3D dielectric cuboids. *Applied Physics Letters, 105*, 084102.

Chapter 2
3D Diffractive Lenses to Overcome the 3D Abby Diffraction Limit

Abstract Radiation cannot be focused on anything smaller than its half of wavelength—or so says more than a century of physics wisdom. In the first part of this chapter the results of the focal fields of a phase correcting Fresnel lens examination are described for several small values of *F*/, with $F \leq 2\lambda$ which allows for overcoming Abbe barrier. It was also shown that the minimum diameter of the focal spot near the central circumferential step of binary diffractive axicon was equal to FWHM = 0.38λ. In the second part of this chapter the innovative radiating structures as a conical FZP lens are proposed for subwavelength focusing. It has been shown that in contrast to the flat diffractive optics the curvilinear 3D diffractive conical optics allows for overcoming 3D Abbe barrier with focal distance F more than F > 2λ.

Keywords Abbe barrier · Diffractive lens · Superresolution · Axicon · 3D optics · Near field

Introduction

Many attempts have made to improve the resolving power of optical imaging systems since Ernst Abbe discovered that the resolution of an imaging system is limited by diffraction. Diffraction, as a general wave phenomenon which occurs whenever a traveling wave front encounters and propagates past an obstruction, was first referenced in the work of Leonardo da Vinci in the 1400s [1] and has being accurately described since the Jesuit astronomer, mathematician and physicist Francesco Grimaldi in the 1600s [1] and by the Scottish mathematician J. Gregory in 1673,[1] who studied the dispersion of white light into a spectrum using a feather. Wave diffraction is a phenomenon that manifests itself as a departure from the laws of geometrical optics under wave propagation.

The erratum to this chapter is available at DOI 10.1007/978-3-319-24253-8_7

---

[1]Rigaud [2].

I. Minin and O. Minin, *Diffractive Optics and Nanophotonics*, SpringerBriefs in Physics, DOI 10.1007/978-3-319-24253-8_2

The possibility to obtain subwavelength resolution utilizes a highly focused near-field spot of a hemispherical lens with a high refractive index in millimeter wave was demonstrated at [3]. In [4] authors demonstrate that in the near-field of a hemispherical sapphire lens a resolution of $\sim 0.3\lambda$ can be obtained for millimeter-wave frequencies. It was mentioned that the first Teflon™ lens with a focal ratio F/D ~ 1, focuses the beam to a diffraction-limited spot. Without a hemispheric sapphire lens the spot size would be approximately one wavelength.

For optical and quasi-optical applications, lenses are typically designed with values for *F*/*D* much larger than one, due to the physically small values for the wavelength. It could be noted due to smaller wavelength, mm-wave lens antennas and lenses rely heavily on the quasioptical principles.[2] These lenses thus have spatial resolutions much greater than a wavelength. At microwave, millimetre-wave and THz frequencies, since the wavelengths are much longer, as well as in microoptics and nanooptics, it is feasible to design lenses with values of $F/D < 0.5$. For these values, the equation $\Delta = 0.61\lambda/\mathrm{NA}$ is no longer an accurate estimate of the resolution. It was not clear what spatial resolution can be obtained for such designs. Moreover, for small values of *F*, the focal spot is in the reactive near field of the lens, and the effects of these reactive field components cannot be neglected, as is the case for larger values of *F* [5].

A new study now shows that it is possible to focus radiation less than half of wavelength, if radiation is focused extremely close to a special kind of "superlens". Diffractive optical elements (DOE)—flat analog of a lens—zone plates (ZP[3]), was a simplest example of DOE. Next, transparent phase gratings and zone plates with a diffractive microrelief were invented, in which the transition from dark to light regions corresponded to a stepwise jump by a quality of the order of wavelength λ, ensuring a wave phase incursion for π.

The values of $F \leq \lambda$ are of interest because the subwavelength resolution may be observed. For example, according to Heisenberg's uncertainty principle, it is impossible to determine the product of the uncertainties of the position of a particle and its momentum to better than h (actually $h/2\pi$): $(\Delta x)(\Delta p_x) > h$ and in scattering of photons in the maximum range of angles, $\Delta x \geq \lambda/2$. However, if we restrict one of the components of the wave vector, it will vary the other components of the wave vector. For example, let $k_y = 0$, $k_z = -i\gamma$, where γ—a positive real number. Than if $\gamma \rightarrow 0$, the range of permissible values of k_x grows indefinitely, and Δx can be arbitrarily small. Imaginary values of k_z correspond to damped waves.

[2]In "quasi-optical systems" the diffraction effects are inevitably important because of although radiation is typically propagated and analyzed as free-space beams, unlike traditional optics, in MMW and THz beams may be only a few wavelengths in diameter. See [5].

[3]The classical Fresnel zone plate, consisting of a plane array of alternately opaque and transparent concentric circular rings, acts upon a normally incident plane wave, transforming it into a converging wave and concentrating the radiation in a small region about a point on the axis. The zone plate is an image forming device, but the mechanism involved for this simple screen is not refraction at the boundary between different dielectric media, but diffraction at the series of annular apertures and subsequent interference of the diffracted radiation.

Consequently, the implementation of subwavelength resolution antenna probe must be within the evanescent field near the surface of the sample, that is certainly at $Z < \lambda$ [6].

It should be noted that *diffractive focusing elements* are crucial components of major range communications and instruments nano systems. Their design and manufacturing would be much faster and cheaper if one could avoid costly prototyping and measurements. To minimize design delays as well as to propose innovative radiating structures, it is necessary *to develop a scale model.* So investigations in millimeter wave/THz taking into account the scale effect may be transfer directly to optical and nanooptical bands.

Flat Diffractive Lens with Superresolution

The binary phase-correcting Fresnel lenses [7, 8] were consist of a set of annular dielectric rings whose radii were determined using the traditional Fresnel zoning rule for flat surfaces with the geometric optics approximation:

$$r_i = \sqrt{Fi\lambda + \left(\frac{i\lambda}{2}\right)^2}, \quad i = 1, 2, \ldots N \tag{2.1}$$

where F is the focal length of the lens, λ is the wavelength, r_i is the radius of the ith Fresnel zone, and N is the total number of zones in the lens of diameter D. (It should be noted that the optimisation of the Fresnel zone radii [9] for improved focusing was not the aim of this study). The thickness (t) of the rings is calculated based on the requirement of achieving a 180° path difference between the waves travelling through the dielectric rings and those travelling through air:

$$t = \lambda / [2(\sqrt{\varepsilon_r} - 1)] \tag{2.2}$$

where ε_r is the dielectric constant of the rings. Three lenses were designed using a dielectric constant of $\varepsilon_r = 4$, which from (2.2) result in a ring thickness of $t = \lambda/2$. The focal lengths for the lenses were chosen to be 2λ, 1λ, and 0.5λ. The diameters for the three lenses are somewhat different since an integer number of zones was used for each case; the values were chosen as close to $D/\lambda = 10$ as possible.

The focusing behaviour for the case of an incident plane wave, polarized in the y-direction, was investigated using finite difference time domain (FDTD) analysis[4] [9–14].

The simulated focal fields in the y-z plane (at $x = 0$) are shown in Fig. 2.1, where the focal spots are seen to the right of the lenses. For all three lenses, the maximum

[4]FDTD simulation of flat FZP was developed in cooperation with N. Gagnon and A. Petosa from Communications Research Centre, Canada.

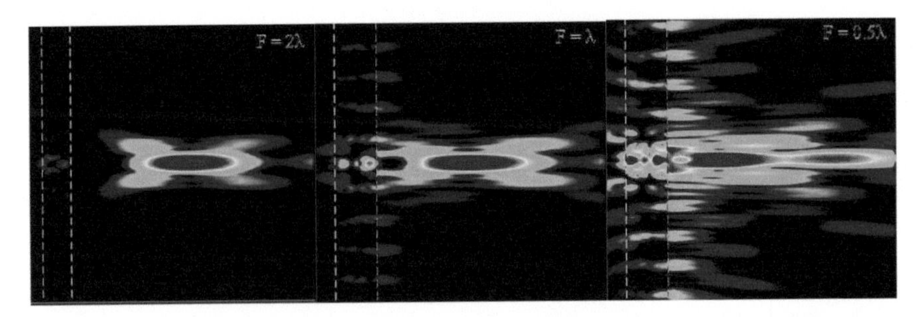

Fig. 2.1 Normalized power density in the y-z plane for the three Fresnel lens designs

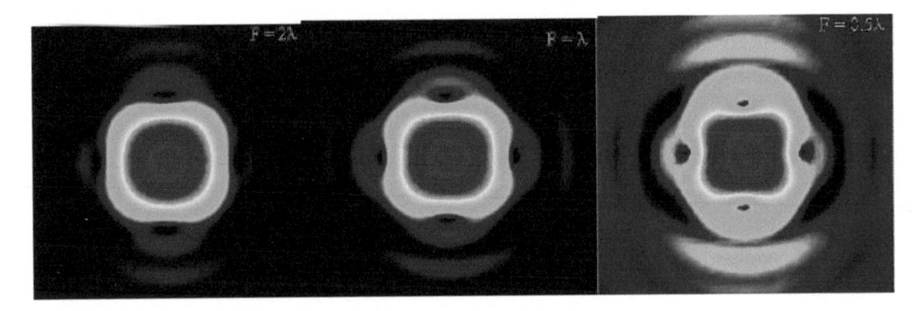

Fig. 2.2 Normalized power density at the distance of maximum intensity (focal spot)

intensity occurs close to the designed focal length. This can be better observed in Fig. 2.2, which plots the power density (normalized to the peak value) along the focal axis ($x = y = 0$). The results of experimental verifications [6] confirm all FDTD simulation results. Since the spot beams were not perfectly axially symmetrical, the spatial resolutions in both the *x*- and *y*-planes are listed (both FWHM and determinated by the 1st minimum). Except for one case ($F = 2\lambda$), the spatial resolutions are all less than 0.5λ, which is significantly finer than the spatial resolution achieved from lenses with large values of *F*/*D*.

This investigation on the focusing properties of phase correcting Fresnel lenses with values of $F/D < 0.2$ and with $F \leq 2\lambda$ has shown that a spatial resolution of less than 0.5λ is achievable. Also from the Fig. 2.2 it is followed that except for one case ($F = 2\lambda$), the axial resolutions are all less than 2.0λ, which is significantly finer than the axial resolution achieved from lenses with large values of *F*/*D*. So the "Abbe barrier" was thus completely broken by such diffractive lenses with unique 3D super resolution [6, 14, 15]. As the *F*/*D* values decrease, so does the spot beam size and thus a finer resolution is achieved. Although the spot beam decreases with decreasing *F*/*D*, the total amount of focusing power also decreases (Fig. 2.3). There is thus a trade off between refined resolution and focusing power, and the selection of the *F*/*D* will thus depend on the intended application.

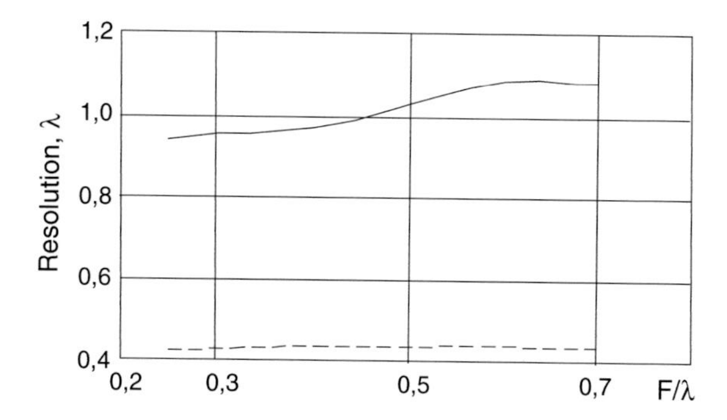

Fig. 2.3 Average resolving power versus F/λ for x-axis (*solid line*) and y-axis (*dashed line*)

The reason for the slight axial asymmetry in the intensity patterns arises from the anti-symmetrical component of the electric field in the z-direction (direction of the incident wave). This component is significant for small values of *F/D*, since the focal spot is in the reactive near-field of the lens, and it causes the slight asymmetry in the intensity patterns (Fig. 2.3). As the values for *F/D* increase, the amplitude of this component quickly decreases and thus does not contribute significantly to the intensity, resulting in a symmetric pattern. In this near field ($F \sim \lambda$) we may talk about evanescent waves, which are waves that are non-homogenous, non-propagating, and exponentially decaying away from the object's surface. They have a complex wave vector perpendicular to the surface and thus allow high wave vectors in the other two directions (i.e. in the plane of the sample). Large k-vectors correspond to small dimensions in direct space, and small directions imply high resolution. Due to the presence of this significant amount of Z polarised light, in the form of a dumbbell along the X axis, the resultant focal intensity distribution is not circularly symmetric [16].

Because of the focal spot in the high NA cases is no longer an Airy pattern and have no circular symmetry it is needed to offer a new criterion of 3D resolution in this case.

Thus in the papers [14, 15] at the first time it were demonstrated that in the near field of single diffractive lens without of the immersion medium a resolution of 0.3–0.4λ can be obtained and the focal spot is slight asymmetry and is no longer an Airy pattern. Moreover, compared to well-known scanning microscopy, this diffractive lens allows simultaneous imaging of a finite area close to the focus.

Later in [17, 18] these results were confirmed in optical wavebands.[5] For example, in [17] numerically and experimentally the focusing of linearly polarized light using the Fresnel ZP with a focal distance of 0.79λ was studied. An elliptical focus spot was experimentally observed with the least diameter by the intensity decay halftime FWHM = 0.63λ.

[5] Dr. Rakesh G. Mote has not come across papers [14, 15]. R.G. Mote, private communication, 2015.

Five years later researches in the optical range [19] also were fully confirmed the results of earlier studies in the millimeter range [14, 15]. In [20] also it was shown that the ability of overcoming the diffraction limit using a zone plate with radius R = 20λ, F = λ and R-TEM 01 mode with radius ω = 10λ is possible. In [21] it was show that the near-field subwavelength focusing can be obtained: with the parameters of the FZP as D = 0.6 μm and F = 0.5 μm (F = 0.79λ), the FWHM was in the range of 0.371λ–0.374λ. The symmetry of focal spot was not investigated.

Thus a high linear resolution may be achieved by using a Fresnel (or Soret) zone plate with NA greater than almost one. The focusing element (lens) can employ also a modified zone plate with an opaque central part which makes up not less than about 50 % of the entire surface area of the plate, in which the radii of Fresnel zones are calculated by taking into account the reference phase concept [7, 22, 23]. But in this case the apodization effects are large.

The principle of the combined zone plate with high NA based on the arbitrary linear phase shift of the radiation passed through one section relative to the radiation passing through the other section also allow to increase the resolution in focal spot. Due to interference effects the sum of the first diffraction orders of the radiation passed through the various sections, the phase shift provides amplification of the longitudinal components of the electromagnetic radiation provided perpendicularity of lines dividing sections of the plane of polarization of the incident radiation. The longitudinal component of the electromagnetic field for FZP with high NA is much stronger and more localized on the optical axis than transverse components of the electromagnetic field. So the introduction of a linear phase jump perpendicular to the direction of linear polarization leads to exclusion in the center of the focal region of the transverse components and the appearance of the longitudinal component of the electromagnetic field. Thus, the prevalence of longitudinal components leads to the reduction of the diameter of the focal spot up to subwavelength size and also decreases the asymmetry of the focal spot due to the attenuation of the transverse component of the electromagnetic field. For the diffractive optical elements (FZP) the introduction of linear phase shift it is possible to realized by means of so-called reference phase concept [7, 22, 23].

The simulation results of photon sieve FZP investigation in millimeter wave had shown [12] that the resolution power of photon sieve FZP is the same as for classical FZP. The reason is on the small number of holes in the FZP aperture. But the first side lobes suppressed conveniently. So the idea of the photon sieve application to millimeter wave optics does not allow increasing the resolution power. But as it known nano-optics deals with optical effects occurring if light interacts with matter that has artificially structured features with sizes comparable to the wavelength.[6] From this point of view, we detailed [12] how the use of simple scale computational experiments in millimeter wave [9] allows obtaining insight into physical systems which are characterized by nanometric objects because of the D/F and D/λ are almost the same.

[6]Novotny [24].

Subwavelength Focusing with Binary Axicon

Demand for improved resolution has stimulated research into developing methods to image beyond the diffraction limit based on different types of optical element including diffractive axicon [7]. As it well known from the Bessel beam theory, the radial distribution of such beam does not depend on the propagation distance z. The effective width of the central peak can be extremely narrow (approximately 3λ/4 far from axicon surface) over large distances.

In millimeter wave/THz if we use a point-like of radiation source in the geometrical optic approximation the radii of zones in the flat axicon surface may be determinates as follows [25]:

$$\sqrt{F^2 + r_k^2} + r_k \sin\gamma = \sqrt{F^2 + a^2} + a \sin\gamma + k\lambda,$$

where k = 1, 2,…, N, $r_N = D/2$, γ—the angle of rays which crossed the optical axis, a—additional free parameter so-called reference radii [26]—see Fig. 2.4.

So

$r_k = -c\frac{\sin\gamma}{\cos^2\gamma} + \frac{1}{\cos^2\gamma}\sqrt{c^2 - F^2\cos^2\gamma}$, where

$c = \sqrt{F^2 + a^2} + a\sin\gamma + k\lambda.$

The resulting expression shows that in contrast to plane wave illuminating the period of such zones in the diffractive element is not constant, but varies along the radius.

To simplify the problem let's consider axicon with plane wave illumination and with reference radius equal to zero. Binary axicon had a height of step as H = λ/2 (n − 1), the index of refraction n = 1.46. The width of the steps were d = 1.2λ, the period was T = 2d = 2.4λ. The diameter of the axicon was 5 periods D = 12λ. The frequency was equal to 100 GHz. The FDTD simulation showed (Fig. 2.5) that with increasing wavelength λ the focal spot is formed closer to the top of the axicon surface, and its diameter in wavelengths decreases [27, 28]. Also studies have shown that at small distances from the axicon surface (z < 10 wavelength) observed the ellipticity of focal spots (eccentricity of the ellipse about 0.63–0.65), which extended along the linear polarization vector of the incident wave on the axicon.

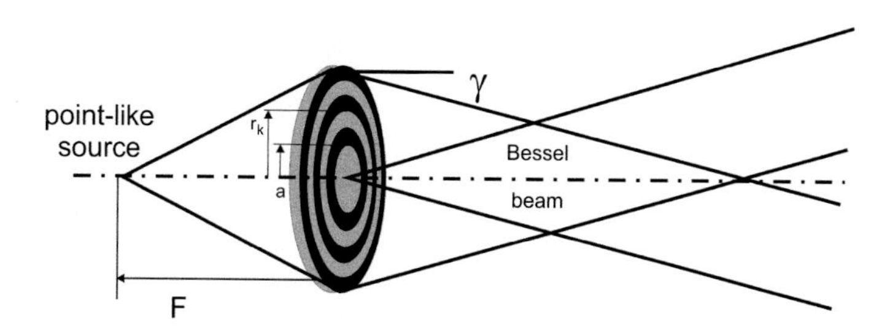

Fig. 2.4 Definition of the binary axicon structure

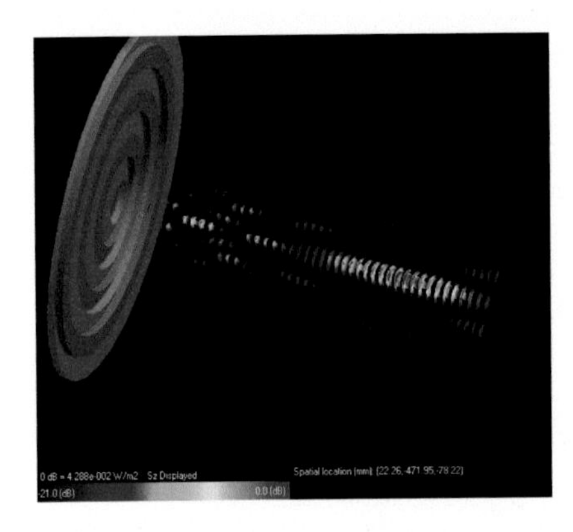

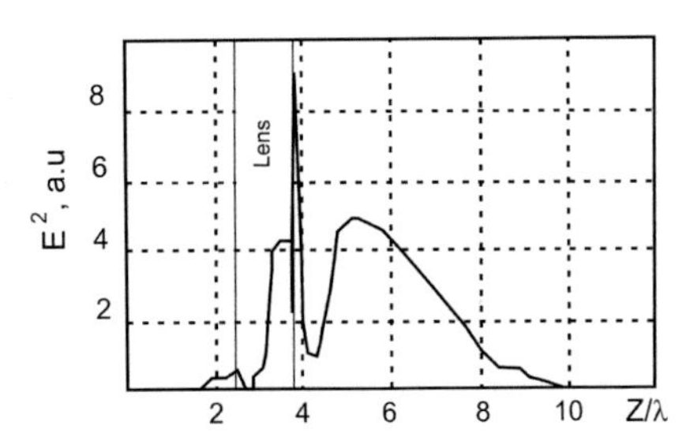

Fig. 2.5 FDTD simulation of axicon focusing (*left*) and field intensity distribution $|E|^2$ along optical axis of optimized binary diffractive axicon (*right*)

The minimum diameter of the focal spot near the central circumferential step was equal to FWHM = 0.38λ and formed at a distance from the axicon surface, about half of wavelength of radiation. Although the present study is at the THz frequency, the principle can be extended to the optical and nanooptical region straightforwardly.

Zoned Metamaterial Lens

A 1.5λ-thick planoconcave zoned lens based on the fishnet metamaterial was investigated experimentally at millimeter wavelengths in [29].The lens was designed to operate at λ = 5.29 mm, where it behaves as an effective negative refractive index medium n = −0.25, with a focal length of F = 9λ. The depth of focus, defined as FWHM along z-axis at the focal length, obtained from experiment was 3.24λ. The sidelobes levels were about 15 %. The symmetry of focal spot was not investigated. To reduce a sidelobes levels the principle of reference phase was successfully apply to the cylindrical fishnet metalenses designed with a $F = 4.5\lambda_0$ and a value of focal length to lateral size ratio (*F/D*) of $F/D_{xz} = 0.214$ and $F/D_{yz} = 0.207$ in the *xz*- and *yz*- planes, respectively (where D_{xz} and D_{yz} are the lateral size of the lens in the *xz*- and *yz* planes, respectively) at [30].

3D Diffractive Conical Lens

Subwavelength resolution beyond the Abbe barrier described above is possible for flat diffractive lens (DOE) only with $F < \lambda$. Below we describe the unique possibilities of 3D subwavelength resolution focusing for 3D diffractive lens with $F > \lambda$.

It is well known [7, 22], that a Fresnel zone plate can be produced conformable to some curvilinear formations. And curvilinear lens can be made on an arbitrary-shaped 3D surface, but the FZP-like lens with a rotational symmetry surface has better radiation characteristics and not only the phase function, but also the 3D surface shape are free parameters that can be used for focusing characteristics optimization, including resolution power both for operating with quasi-monochromatic radiations and femtosecond pulses [31, 32]. The advantages of 3D FZP lens are they have more levels of design freedom and optimization. For example, the conical FZP lens can be easy manufactured at millimeter wave and/or terahertz frequencies as a multilayer assembly of dielectric rings embedded in air space or solid dielectric.

The innovative radiating structures as conical millimeter wave FZP lens and lens antenna at were first proposed by the authors in 1991 (Fig. 2.6), described and studied both theoretically and experimentally in [7, 22, 33]. Like the plane phase-reversal flat FZP lens, the cone-shape zone plate lens transforms in a step-wise manner the incident plane wave into a spherical wave converged in geometrical-optic approximation to the primary focus F [33].

As it was shown theoretically and experimentally the longitudinal resolving power and depth definition (axial resolution Δ_z) can be controlled by choosing the flexure of the diffractive lens surface [7]. The specific behavior of axial resolution of diffractive optical elements on a non-flat surface [7, 22] makes it possible to design systems that possess much higher 3D resolution power and gain than other known classical lenses. Therefore, the main conclusion is that the longitudinal resolving

СОВЕТСКАЯ РОССИЯ ◆ 8 февраля 1991 г. ◆ № 28 (10479) ◆ 2

Антенна близнецов

Братья Игорь и Олег Минины работают в Институте прикладной физики Сибирского отделения Академии наук СССР. Проблема, которой они занимаются, связана с завтрашним днем нашего спутникового телевидения.

Творческая группа, в составе которой работают братья Минины, подготовила документацию для серийного производства спутниковых антенн из нетрадиционных материалов.

НА СНИМКЕ: научные сотрудники Игорь и Олег Минины.

Фото Б. Евгеньева

Fig. 2.6 Prototypes of the conical, spherical, parabolic and ogival-shape mm-wave antennas 1991

power of the diffractive optical element can be controlled by choosing the flexure of the diffractive optical element surface and its spatial orientation [7, 8]. Also the latter important effect is due to the reduced FZ lens spherical aberration and to the reduction of the zone shadowing effect in diffractive optical elements on curvilinear surfaces [7, 31]. It could be noted that by selection of diffractive optic surface and its orientation it is possible to minimize the selective types of aberrations [34].

Results of Investigations

Some details about the experimental setup and experimental technique can be found at [7, 35]. First, as an example, we briefly consider a diffractive optical element fabricated on a conical surface which was studied by the authors in 1983 [7, 22]. It was manufactured of a numerically controlled lathe, using optical-grade polystyrene with the following optical constants: diffractive index n = 1.59 and absorption coefficient k $\sim$ 10^{-3}. The nominal radiation wavelength was λ_0 = 4.6 mm, lens aperture $D/\lambda_0 = 44$ and the rear segment F = D (lens factor F/D = 1). The maximum flexure of the diffractive lens surface was $\langle x \rangle/\lambda = 32$. The Fresnel number of lens is 138.

An analysis of the results [36] obtained shows the following:

- half-width (at half-height) of field intensity distribution along the optical axis for a "conical" diffractive optical element, with parameters as shown above, is twice as narrow as that of an equivalent zone plate (when radiation is incident on the side of the apex of the diffractive optics);
- when radiation is incident on the side of the base of the DOE, the width of field intensity distribution along the optical axis is approximately 2.5 times wider than for the equivalent zone plate;
- the shape of this distribution, plotted in relative units, varies very little (by about 3 %) in the range of wavelengths that deviate from the nominal value by less than ±17 %;
- as wavelength decreases in comparison with the nominal value, the intensity of the first side lobe increases; this is the lobe that is located further from the zone plate relative to the distribution maximum. The amount of increase of relative intensity approximately coincides with the amount of wavelength detuning.

Now, consider briefly the conical diffractive lens with small diameter. The parameters of the diffractive lens (Fig. 2.7) both binary and phase correcting was selected as follows: F/D = 1.26, D/λ = 20 (the dimension of lens diameter was limited by the computer's capabilities), F/λ = 25, $90 < \alpha < 20$, the lens material was polystyrene (n = 1.59). In the case of binary conical FZP the lens consists of several metal screens of different annular hole normal to optical axis and situated along the conical surface [7, 22]. The results of FDTD simulation of different type of conical phase reversal zone plate with different cone angle are listed below in Fig. 2.8.

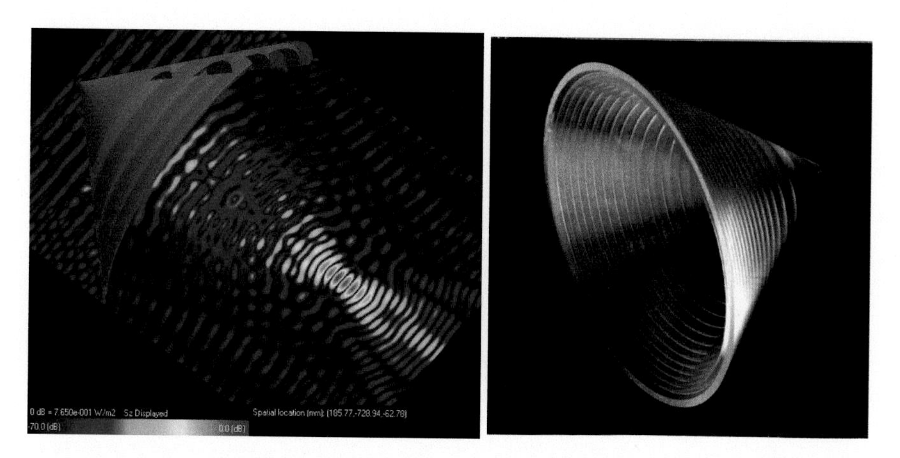

Fig. 2.7 FDTD simulation of conical lens focusing (*left*, *red*—air, *green*—dielectric) and experimental diffractive conical lens (*right*)

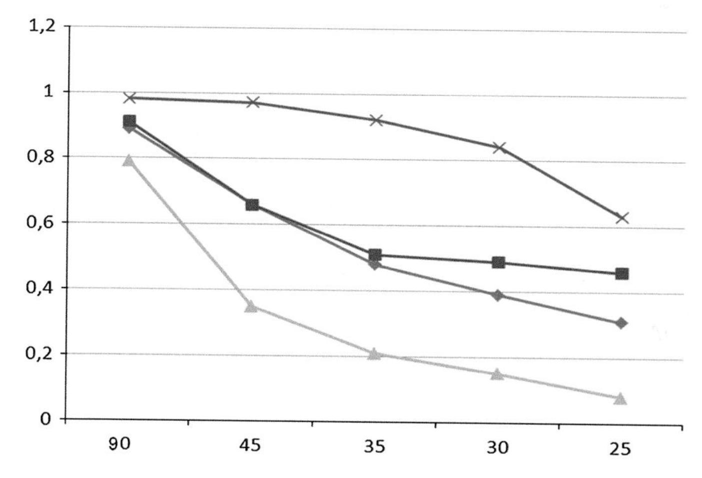

Fig. 2.8 FDTD simulation of resolution power of 3D conical diffractive lenses: *blue*—Δx, *red*—Δy, *green*—Δz, the *purple* curve indicate the asymmetry of focal spot Δx/Δy. The value of Δz is in the unit of classical depth of focus $8\lambda\left(\frac{F}{D}\right)^2$[1]. All other values are in the unit of Rayleigh radius [1]. At the horizontal axis the cone half-opening angle is shown

The results of simulations in the units of wavelength are shown in the Table 2.1. In approximation the focal spot is ellipse with the length of Δx/λ and Δy/λ in the Table the square of focal spot S are shown. Also the "value" of 3D focal spot V in FWHM on axial direction and lenses Gain are shown in the Table 2.1. The comparison of FDTD simulations and simple approximate algorithm [37] has shown a good agreement. Also it is surprising that simple model of diffractive lens based on several flat metallic annular rings placed along conical surface and normal to optical axis [33, 37] gives the results similar to classical dielectric conical diffractive lens.

Table 2.1 Focusing characteristics of dielectric diffractive conical lenses

α (°)	Δx/λ	Δy/λ	Δz/λ	ΔF/λ	S/λ^2,	V/λ^3	G, dB
90	1.31	1.32	9.2	23	1.36	12.49	11.8
45	0.92	0.91	3.65	13	0.65	2.4	21.1
35	0.67	0.70	2.47		0.37	0.91	21.7
30	0.55	0.68	1.7	6.4	0.29	0.5	21.9
25	0.42	0.67	0.90	2.45	0.22	0.2	22.4

In experimental verification we used a diffractive optical element fabricated on a conical surface. It was manufactured of a numerically controlled lathe, using optical-grade polystyrene with the following optical constants: diffractive index n = 1.59 and absorption coefficient k ~ 10^{-3}. The nominal radiation wavelength was λ_0 = 4.6 mm, the lens factor F/D = 1. The initial lens aperture D/λ_0 = 44 from [7] was limited to a value of D/λ_0 = 20 by absorption materials to compare with simulations and cone angle α = 35°.

The analysis of the experimental results shows:

- half-width (at half-height) of field intensity distribution along the optical axis for a "conical" diffractive optical element, with parameters as shown above, is twice as narrow as that of an equivalent zone plate (when radiation is incident on the side of the apex of the diffractive optics) and less than $\Delta_z < 2\lambda$;
- when radiation is incident on the side of the base of the diffractive optical element, the width of field intensity distribution along the optical axis is approximately 2.5 times wider than for the equivalent zone plate [7];
- the resolution power of conical lens is about 0.7 of wavelength with full cone angle of 70° in the first case.

It could be noted that the distance from the base of the cone to the focal point ΔF/λ is always ΔF > 2λ (see Table 2.1). Therefore, the longitudinal resolving power (axial resolution) of the diffractive optical element can be controlled by choosing the flexure of the diffractive optical element surface and its spatial orientation and could be less than Abbe barrier. So the "Abbe barrier" was completely broken by such diffractive lenses with unique 3D super resolution.

So in contrast to the flat diffractive optics the curvilinear 3D diffractive conical optics allows for overcoming 3D Abbe barrier with focal distance F more than F > 2λ. The focal intensity distribution for conical diffractive lens (as for phase reversal flat FZP lens [7, 33]) is also not circularly symmetric and thus the focal spot in the high NA case is no longer an Airy pattern. These results may find useful applications in optical microscopes, including "reverse-microscope", nondestructive testing, microoptics, nanooptics, for manipulate the 3D focused field distribution flexibly by use of diffractive optical elements to some applications and so on.

References

1. Born, M., & Wolf, E. (2005). *Principles of optics*. Oxford: Pergamon.
2. Rigaud, S. J. (ed.). (1841). *Correspondence of scientific men of the seventeenth century* (Vol. 2, pp. 251–255). Oxford: Oxford University Press.
3. Mansfield, S. M., & Kino, G. S. (1990). Solid immersion microscope. *Applied Physics Letters, 57*, 2615–2616.
4. Pimenov, A., Loidl, A. (2003). Focusing of millimeter-wave radiation beyond the Abbe barrier. *Applied Physics Letters, 83*, 4122. doi:10.1063/1.1627474
5. Goldsmith, P. F. (1998). *Quasioptical systems*. New York, IEEE Press
6. Minin, I. V., & Minin, O. V. (2014). Experimental verification 3D subwavelength resolution beyond the diffraction limit with zone plate in millimeter wave. *Microwave and Optical Technology Letters, 56*(10), 2436–2439.
7. Minin, O. V., & Minin, I. V. (2004). *Diffractional optics of millimeter waves*. Boston: IOP Publisher.
8. Minin, I. V., Minin, O. V. (2008). *Basic principles of Fresnel antenna arrays*. Lecture Notes Electrical Engineering (Vol. 19). Berlin, Springer.
9. Minin, I. V., Minin, O. V. (2003). Technology of computational experiment and mathematical modeling of diffractive optical elements of millimeter and submillimeter waveband. In *International Conference "Information Systems and Technologies" IST-2003*, Novosibirsk, NSTU (vol. 1, p. 124–130), April 22–26, 2003.
10. Remcom Incorporated. http://www.remcom.com/html/fdtd.html
11. Yee, K.S. (2010). Numerical solution of initial boundary value problems involving Maxwell's equations in isotropic media. In *IEEE Transaction on AP-14*, 1966, No. 3, pp. 302–307. See also: John B. Schneider. Understanding the Finite-Difference Time-Domain Method 2010. www.eecs.wsu.edu/~schneidj/ufdtd
12. Minin, I. V., & Minin, O. V. (2013). FDTD analysis of millimeter wave binary photon sieve Fresnel zone plate. *Open Journal of Antennas and Propagation, 1*(3), 44–48.
13. Iwata, S., Kitamura, T. (2011). Three dimensional FDTD analysis of near-field optical disk. In *Progress in Electromagnetics Research Symposium Proceedings* (pp. 157–160). Marrakesh, Morocco, March 20–23, 2001.
14. Minin, I. V., Minin, O. V., Gagnon, N., Petosa, A. (2006). FDTD analysis of a flat diffractive optics with sub-Reyleigh limit resolution in MM/THz waveband. In *Digest of the Joint 31st International Conference on Infrared and Millimeter Waves and 14th International Conference on Terahertz Electronics* (p. 170). Shanghai, China, September 18–22, 2006.
15. Minin, I. V., Minin, O. V., Gagnon, N., Petosa, A. (2007). Investigation of the resolution of phase correcting Fresnel lenses with small focal length-to-diameter ratio and subwavelength focus. In *EMTS*, Canada, Ottawa (URSI), July 26–28, 2007. See also: Minin, I. V., Minin, O. V., Gagnon, N., Petosa, A. Investigation of the resolution of phase correcting Fresnel lenses with small values of F/D and subwavelength focus http://www.computeroptics.smr.ru/KO/PDF/KO30/KO30111.pdf
16. Theory of high numerical aperture focusing. http://www.iitg.ernet.in/physics/fac/brboruah/htmls/hnaf.html
17. Mote, R. G., Yu, R. G., Kumar, A., et al. (2011). Experimental demonstration of near field focusing of a phase micro Fresnel zone plate (FZP) under linearly polarized illumination. *Appl. Phys. B, 102*, 95.
18. Mote, R. G., Yu, S. F., Zhou, W., et al. (2009). Subwavelength focusing behavior of high numerical aperture phase Fresnel zone plates under various polarization states. *Applied Physics Letters, 95*, 191113.
19. Stafeev, S. S., O'Faolain, L., Shanina, M. I., Kotlyar, V. V., Soifer, V. A. (2011). Subwavelength focusing using Fresnel zone plate with focal length of 532 nm. *Computer optics, 35*(4), 460–461 (2011) (in Russian). See also: Kotlyar, V. V., Stafeev, S. S., Liu, Y.,

O'Faolain, L., Kovalev, A. A. (2013). Analysis of the shape of a subwavelength focal spot for the linearly polarized light. *Applied Optics, 52*(3), 330–339.
20. Stafeev, S. S., Kotlyar, V. V. Comparative modeling two methods of sharp focusing with zone plate. *Computer optics, 35*(3), 305–310 (in Russian).
21. Zhang, Y., Zheng, Ch., Zhuang, Y., Ruan, X. (2014). Analysis of near-field subwavelength focusing of hybrid amplitude–phase Fresnel zone plates under radially polarized illumination. *Journal of Optics, 16*, 1-6 015703
22. Minin, I. V., Minin, O. V. (1992). *Diffractive optics* (180 p). NPO InformTEI, Moscow (in Russian).
23. Minin, I. V., Minin O. V. (2011). Reference phase in diffractive lens antennas: a review. *Journal of infrared, millimetre and THz waves, 32*(6), 801–822.
24. Novotny, L., Hecht, B., *Principles of nano-optics*. Cambridge: Cambridge University Press.
25. Minin, I. V., Minin, O. V. (September, 2003) Scanning properties of the diffractive "LENS-PLUS-AXICON" lens in THz, in *Proceedings of the 11th Microcol Symposium*, (pp. 233–236). Budapest, Hungary, September 10–11, 2003.
26. Minin I. V., Minin, O. V. (1989). Optimization of focusing properties of diffraction elements. *Soviet Letters to the Journal of Technical Physics, 15*(23), 29–33.
27. Minin I. V., Minin, O. V. (2004). Scanning properties of diffractive element forming the axial —symmetric diffraction limited wave beam. *Computer optics, 26*, 65–67 (In Russian).
28. Minin, I. V., & Minin, O. V. (2013). Active MMW/terahertz security system based on bessel beams. *ISRN Opticls, 2013*(285127), 1–4.
29. Pacheco-Pena, V., Orazbayev, B., Torres, V., Beruete, M., & Navarro-Cia, M. (2013). Ultra-compact planoconcave zoned metallic lens based on the fishnet metamaterial. *Applied Physics Letters, 103*, 183507.
30. Pacheco-Pena, V., Navarro-Cia, M., Orazbayev, B., Minin, I. V., Minin, O. V., Beruete, M. (2015). Zoned fishnet lens antenna with optimal reference phase for side lobe reduction. *IEEE Transactions on Antennas & Propagation* (accepted).
31. Minin, I. V., Minin, O. V. (2004). Reduction of the zone shadowing effect in diffractive optical elements on curvilinear surfaces. *Optoelectronics, instrumentation and data processing, 40*(3) (2004).
32. Minin, I. V., Minin, O. V. (2004). Correction of dispersion distortion of femtosecond pulses by using the non-planar surface of diffractive optical elements. *Chinese Optics Letters, 08*, 1–4 (2004).
33. Minin, I. V., & Minin, O. V. (2014). 3D diffractive lenses to overcome the 3D Abbe subwavelength diffraction limit. *Chinese Optics Letters, 12*, 060014.
34. Minin, I. V., Minin, O. V. (2004). Correction of dispersion distortion of femtocesond pulses by choosing the surface shape of diffractive optical elements. *Optoelectronics, Instrumentation and Data Processing, 40*(1), 34–38 (2004).
35. Kim, W.-G., Thakur, J. P., & Kim, Y. H. (2010). Efficient DRW antenna for quasi-optics feed in W-band imaging radiometer system. *Microwave and Optical Technical Letters, 52*, 1221.
36. Minin, I. V., Minin O. V. (2014). Spectral properties of 3D diffractive lenses with 3D subwavelengh focusing spot. In *Proceedings of the 12th International Conference on Actual Problems of Electronics Instrument Engineering (APEIE)-34006*(Vol. 1, pp. 485–487). Novosibirsk, October 2–4, 2014.
37. Minin, I. V., & Minin, O. V. (2001). New possibilities of diffractional quasioptics. *Computer optics, 22*, 99. (in Russian).

Chapter 3
Subwavelength Focusing Properties of Diffractive Photonic Crystal Lens

Abstract Subwavelength focusing properties of diffractive photonic crystal lens was considered. It was shown that photonic crystal lens design has not an unique solution and at least of three different types of photonic crystal lens are possible with spatial resolution at focus equal to FWHM = 0.48λ. PhC diffractive lens with mode transformation are also discussed. It was shown the lens focuses radiations to a spot smaller than the diffraction limit with FWHM = 0.35λ. In the second part we suggest a metamaterial based structure whose properties are determined not only by its inner geometry but also by its entire 3D shape. The example of metacuboid are described. We evaluate the potential of this structure to control both the size and the location of the field enhancement (photonic jet). Effect of EM strong localization in photonic crystal is described.

Keywords Photonic crystal · Subwavelength focusing · Metamaterial · Metacuboid · Strong localization · Bloch theory

Introduction

As it was mentioned one of the factors that have stimulated much of the recent interest in diffractive optics at any frequency waveband has been the increased optical performance of such optical elements. This allows the fabrication of optical elements that are smaller (compared to wavelength), lighter and cheaper to fabricate, are more rugged and have superior performance that the conventional optical or/and quasioptical components they often replace. Important, the design capabilities for binary optics now available can make possible the design and manufacture to components including antennas having optical and focusing properties never before produced. Fresnel zone plate lens (FZPL) is one of the simple digital lenses [1]. Flat surfaces are two dimensional, therefore much cheaper to fabricate than three dimensional contour surfaces and allow to focusing of a radiation to a subwavelength focus distances [2] (see Chap. 2).

I. Minin and O. Minin, *Diffractive Optics and Nanophotonics*,
SpringerBriefs in Physics, DOI 10.1007/978-3-319-24253-8_3

A photonic crystal is a periodic dielectric structure with lattice spacing of the order of the wavelength of the electromagnetic wave. Typical for a photonic crystal is that electromagnetic waves in a certain frequency range and/or with a certain polarization cannot propagate along certain directions in the crystal. This forbidden frequency range is called a stopgap. If the propagation of the electromagnetic wave is forbidden for any crystalline direction and any polarization, for a certain frequency range, then this forbidden frequency range is called a photonic band gap [3, 4].

The idea of controlling radiations by means of photonic crystals[1] has led to many proposals for novel devices [3, 6, 7] and has motivated many researchers to investigate a plethora of ideas. Theoretical possibility to overcome the diffraction limit includes specially constructed photonic crystals and a metamaterial with negative refractive index, for which the superlens effect has been demonstrated [8]. For example, for the superlens [8], the limiting value of the focal spot is described by the Bessel function $J_0(kr)$ and gives the value of the diameter at half intensity FWHM = 0.35λ.

Daschner et al. [9] described a concave lens based upon a photonic crystal that have an effective index of refraction $n_{eff} < 1$. It was shown that the concave lens made from dielectric rods (Al_2O_3) can overcome the diffraction limit at least for the TE polarization; while in the TM polarization it reaches this limit. The lens has the size of 13… 29λ and focal length about diameter (13.5… 30λ), depending on the measured wavelength (from $\lambda = 2.5$ to 5.5 cm). The focal spot area was defined as FWHM. The smallest ratio of the measured focusing area was $0.24\lambda^2$ for the TE polarization, (which is well beyond the focusing limits of generic lenses known from linear optics) and $1.02\lambda^2$ for the TM polarization. It also was mentioned that in the range between $\lambda = 3.1$ and $\lambda = 3.6$ cm (TM) the focusing point moves along the symmetry axis.

Below, based on [10–13], we shown the possibility of subwavelength focus diffractive photonic crystal lens as a perspective focusing element and it has been shown that diffractive photonic crystal lens designs have not an unique solution. Lens diameter was in 5 times more than her width and full width half maximum diameter of focal spot was 0.48λ. As for axial resolution determinate to FWHM it was shown that for the diffractive photonic these values equal to 0.7λ. It was also shown that the diffractive photonic crystal(PhC) lens located at the output of the waveguide focuses radiation to a spot smaller than the diffraction limit.

Numerical simulation has been conducted by using FDTD-method [14]. Input of plane wave with TE-polarization into a computation area is achieved with "total field-scattered field" technique [15]. PML boundary conditions were used [16]. Averaging of field has been performed by one period [11].

[1]It could be noted that a partial lensing effect, light concentration resembling a photonic jet (see Chap. 4), at some conditions may be obtained in PhC lens, as it was noted by Antonakakis et al. [5].

Photonic Crystal Diffractive Lens

The optical length in the PhC lens was calculated as follows[2] [11]:

$$\Delta = N[2r_i(y) + (a - 2r_i(y))n] \tag{3.1}$$

where: N is the number of holes in line, a is lattice parameter (period), $r_i(y)$—the radius of holes, y is vertical axis. From the (3.1) it is followed that photonic crystal lens design have not an unique solution and at least of three different types of photonic crystal lens are possible. The main types of the lens are:

(1) the first type of lens where N = const
(2) the lens of second type where $r_1 = r_2$ and N $\neq$ const
(3) the lens of third type where $r_1 \neq r_2$ and N $\neq$ const

Let us consider the first type of lens where N = const (Fig. 3.1a). The main parameters of diffractive photonic crystal lens are the following:

- Wavelength λ = 10 mm,
- Width of a lens $l = (0.83 + 1)\lambda = 1.83\lambda$
- Array parameter $a = l/10 = 0.183\lambda$
- Radii of a circular holes: $r_1 = 0.25a = 0.0457\lambda$, $r_2 = r_1 + 0.227a = 0.087\lambda$
- Lens positions at z-axis: (1.5–3.3)λ
- Index of refraction n = 1.6

The geometry and dimensions of the lenses is shown in the Fig. 3.1.

For the lens of second type where $r_1 = r_2$ and N $\neq$ const (Fig. 3.1b) the main parameters of diffractive photonic crystal lens are the following:

- Wavelength λ = 10 mm,
- Width of a lens $l = (0.83 + 1)\lambda = 1.83\lambda$
- Array parameter $a = l/10 = 0.183\lambda$
- Radii of a circular holes: $r = qa_1 = 0.2a = 0.0366\lambda$,
- Array constant $a_2 = a_1/\left(1 + \frac{n-1}{2q}\right) = 0.0732\lambda$
- Lens positions at z-axis: (1.5–3.3)λ
- Index of refraction n = 1.6

In this case the optical length in the lens (3.1) was calculated as:

$$\Delta = N_i(y)[2r + (a - 2r)n] = N_i(y)[2r(1 - n) + an] \tag{3.2}$$

[2]The authors [17] considered only a rectangular PhC Mikaelian lens and derived the air hole radii of each row from the assumption that the optical path length along the line which crosses the holes from the center should be equivalent to that of the ideal Mikaelian lens. The main limitation of this approach is that it can be applied only to a rectangular lattice PhC Mikaelian lens with varying air hole radius.

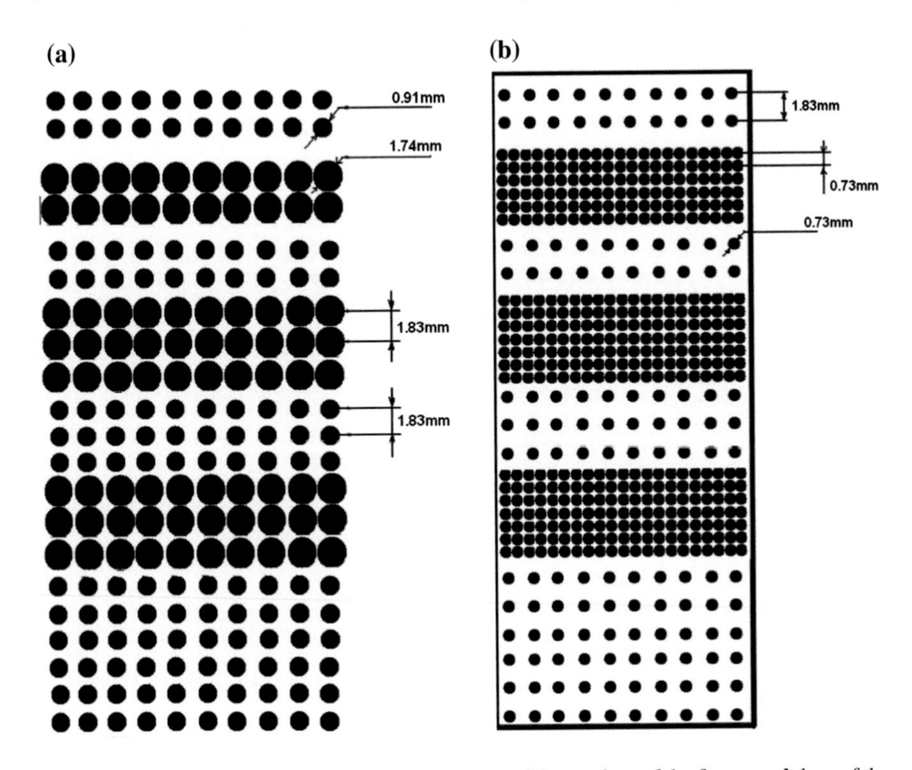

Fig. 3.1 Geometry and dimensions of the diffractive PhC lens: **a** lens of the first type, **b** lens of the second type

Figure 3.2 shows the intensity distribution along (a, b) and across (focal plane) (c, d) optical axis for photonic crystal lens of the first (a, c) and second (b, d) types.

The diameter of beamspot (spatial resolution) at focus was equal to full width half maximum (FWHM): FWHM = 0.48λ.[3]

PhC Diffractive Lens with Mode Transformation

Let us concider the PhC lens of second type (where $r_1 = r_2$ and $N \neq$ const). It was simulated PhC diffractive lens located at the input of the waveguide. Simulations shown that the size of the focal spot is FWHM = 0.37λ, and the longitudinal size of the focus is FWHM = 0.53λ.

PhC lens with the parameters of the previous example was also simulated but PhC lens was located at the output of the waveguide. The simulation results shows

[3]I.V. Minin, O.V. Minin. Diffractive photonic crystal lens of millimetre waves. Patent of Russia 152929.

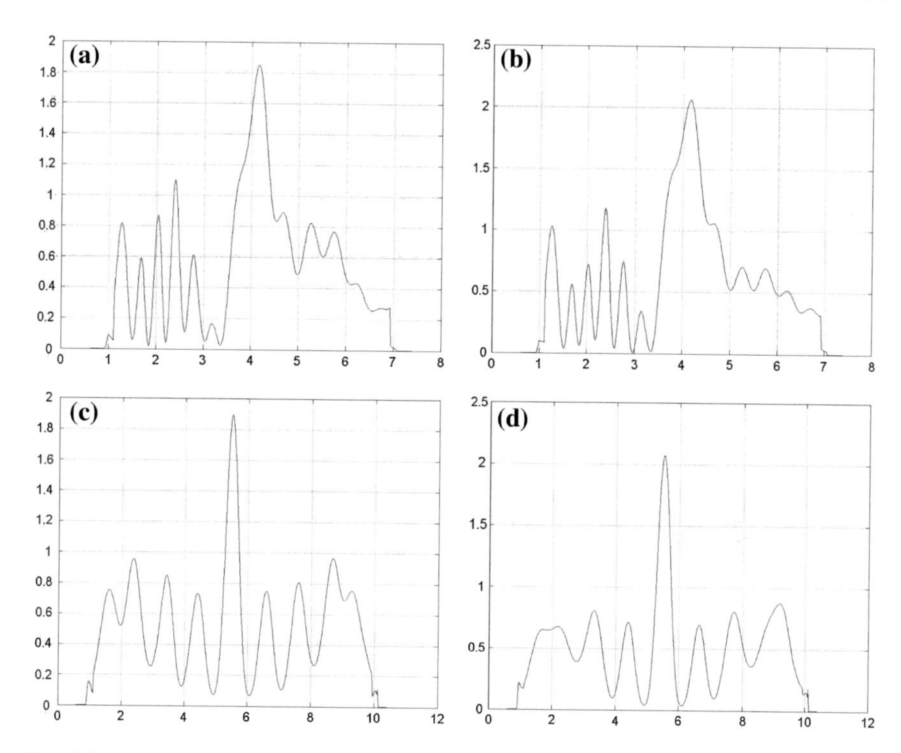

Fig. 3.2 Field intensity distribution along (**a**, **b**) and across ("focal spot") (**c**, **d**) optical axis for photonic crystal lens of the first (**a**, **c**) and second (**b**, **d**) types. The relative units on the x-axis are in wavelength

that the diameter of the focal spot at half intensity is FWHM = 0.314λ. A comparison of these 2 types of lenses shows that in addition to decreasing diameter of the focal spot in the case of PhC lens in the waveguide, the sidelobes of the diffraction pattern at the focus were also smaller. Simulations also shown that at the same time, increasing the number of phase quantization levels to 6–8 allow to increase the intensity of focus to a value greater intensity for the FZP lens in free space.

Important that as it well known the scalar theory in the 2D case describes a diffraction-limited focus by the sinc-function: $Ex(y, z) = \mathrm{sinc}(2\pi y/(\lambda NA))$ which is at a maximum numerical aperture $NA = 1$ gives the diffraction limit of the focal spot with the diameter at half intensity FWHM = 0.44λ. As it was mentioned above for the superlens [8] the value of the spot diameter at half intensity FWHM = 0.35λ. Thus, the lens in described below focuses radiations to a spot smaller than the diffraction limit.

It could be mentioned that the concept of flat lens imaging based on photonic crystal may be also addressed from different approaches such as an anisotropic effect [18], partial band gaps [19], the role of reflection [20], the effect of channeling from self-collimation [21] and on the existence of surface modes [22], etc.

In conclusion of this part we could mention that in contrast to array of holes in [23] sub-wavelength imaging based on array of non-spherical nano-particles was demonstrated. It was shown that in lenses with properly designed layers of anisotropic material, the evanescent waves can be converted to propagating waves and the sub-wavelength imaging process is less sensitive to the material loss of individual constituents.

Metacuboid-Aided Photonic Jet

In this sub-chapter we suggest a metamaterial structure whose properties are determined not only by its inner geometry but also by its entire 3D shape. We evaluate the potential of this structure to control both the size and the location of the field enhancement (photonic jet—see Chap. 4). The idea of metacuboid-aided photonic jet is to take a dielectric cuboid particle, described in Chap. 4, and then realize it using a photonic crystal operating as an effective medium, that is, in the so-called metamaterial regime. This can be done using two dimensional photonic crystal consisting of parallel rods in air host medium, or air cylindrical holes inside dielectric matrix. Both of them can be homogenized and then it is possible to realize the specified cuboid. In that case, cuboid made as a PhC would be polarization dependent, it would work only for the polarization for which effective refractive index is equal to cuboid material.

A dielectric plate with cavities of various shapes is one of man-made dielectric type in which the effective dielectric permittivity is reduced in comparison with a solid dielectric because the fraction of volume that the dielectric occupies is reduced. The perforations result in changing the effective dielectric constant of the dielectric material [24–26]. In Ref. [24] the results of an experimental investigation of a perforated dielectric in the resonance region were discussed. In our preliminary investigations we used a triangular perforation cell [26, 27]. To minimized the losses to reflection we have selected $\alpha \approx 0.4342$ which corresponds to $\varepsilon_r \approx 3$. Taking into account that $\sqrt{\varepsilon_{eff}} = n_{eff} = 1.46$ [28] the parameters of the triangular cell are: $\alpha \approx 0.4342$, $s \approx 0.1445\lambda$, $d \approx 0.1\lambda$. The 3D elementary dielectric metacube with dimension $L \times L \times L$ ($L = \lambda$) is schematically shown in Fig. 3.3. The whole structure is illuminated by using a vertically polarized plane wave (E_y) with its propagation direction along the optical z-axis.

Simulation results are shown in the Fig. 3b. The ellipticity (defined as the ratio between both transversal resolutions $FWHM_x/FWHM_y$) [28] of jet are close to 1. Therefore, a quasi-spherical spot is obtained for the metacuboid configuration. The parameters of a jet are: $FWHM_x = 0.618\lambda$, $FWHM_y = 0.617\lambda$, $FWHM_z = 1.89\lambda$.

For the second variant of metacuboid we have selected the parameters as follows: $\varepsilon_r \approx 2.75\alpha \approx 0.3092$, $d \approx 0.1\lambda$, $s \approx 0.17126\lambda$. The photonic jet parameters are: $FWHM_x = 0.555\lambda$, $FWHM_y = 0.542\lambda$, $FWHM_z = 1.61\lambda$.

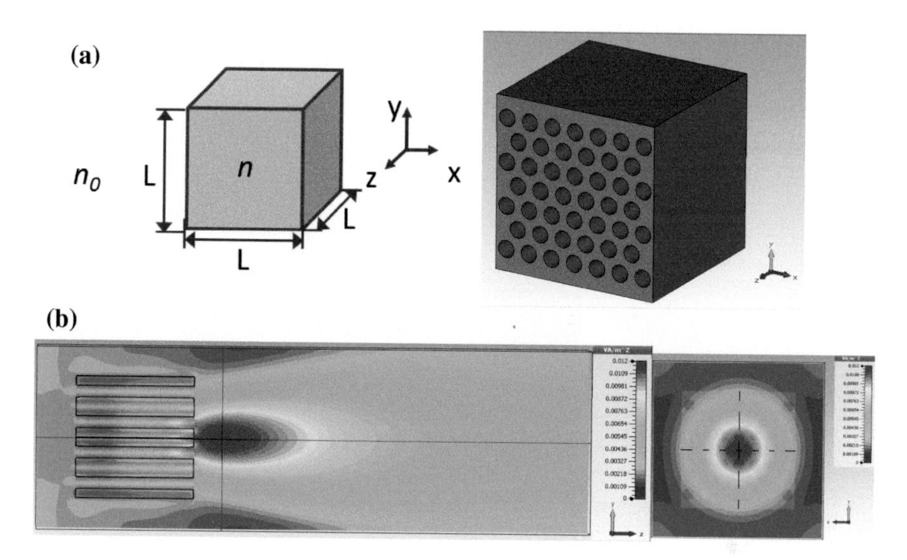

Fig. 3.3 **a** Schematic view of dielectric metacube concept, **b** Simulation of photonic jet by metacuboid

Thus a concept and principal possibilities of photonic jet formation based on metamaterial cuboid is shown. So the photonic crystals in general have served to revisit old and apparently well-established concepts, shading new light on them. They have been the playground where physicists and engineers have worked together and this has produced a very fast development of new ideas and applications. We hope we are opening new research lines such as mesoscale metamaterial focusing devices for sensing applications.

Effect of EM Strong Localization in Photonic Crystal

As it well known photonic crystals (PhCs) are materials which have a periodicity dielectric constant in some particular dimensions [29]. They are artificial multi-dimensional periodic structures with a period of the order of electromagnetic (EM) wavelength [30]. Due to this periodicity, it is usually assumed that the radiation propagates in the form of Bloch modes—it means that if you multiply a plane wave by a periodic function, you get a Bloch wave [29–31]. The existence of spatially periodic fluctuations of medium properties with a period comparable to the wavelength of EM radiation results in the fundamental transformation of the elementary excitations spectrum, the formation of allowed and forbidden photonic bands, etc. [29–32].

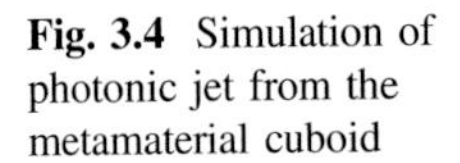

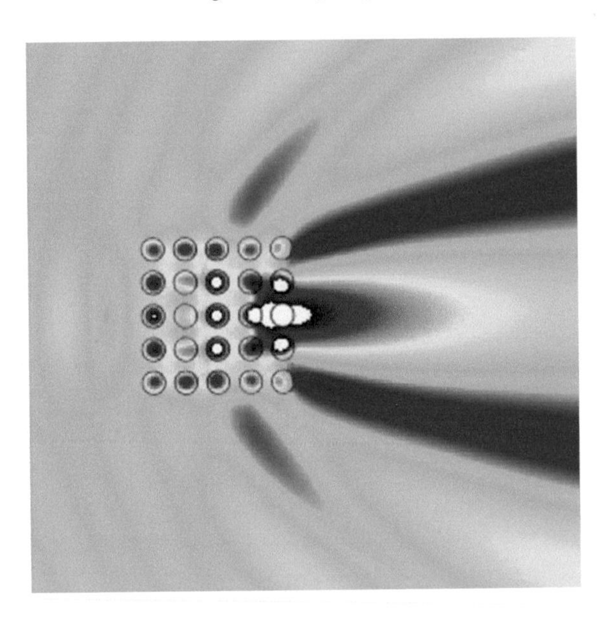

Fig. 3.4 Simulation of photonic jet from the metamaterial cuboid

But near the surface of photonic crystal (air/PhC interface) the strong local localization of incident wavefront (EM) are observed due to the specificity of the interaction of incident radiation with photonic crystal surface (Fig. 3.4) and leads to the strong localization of electric field near the PhC surface structure during pumping in its band gap due to coherent distortion of incident wavefront. It is important to note that this effect can't be predicted or explain by the Bloch theory because of the wavefront is distorted and not a plane along the photonic crystal and so this effect does not allow to described the EM propagation through PhC as the interaction of the plane EM wave with an equivalent 1D structure. These wavefront distortions are determined by the internal structure of the photonic crystal and illuminated wavelength. We call this effect as a EM strong localization (EMSL) in PhC.

Figure 3.4 show the results[4] for the terajet with n = 1.41 [28] and realized using 2D photonic crystals with dielectric rods (eps = 11) in air host medium. One could see the reflection from the structure due to high permittivity of rods.

The simulation of the interaction of EM with the air (n = 1) PhC matrix shown that high-intensity peaks are localized in a small volume of the photonic crystal within PhC cylindrical rods. The peaks can be much more of magnitude than the radiation intensity in a homogeneous dielectric material at the same power of the electromagnetic wave incident on the medium and cannot be predicted from the point of view of the Bloch wave's theory.

It was also could be noted that the structure described above (Fig. 3.4) is *birenfringent*, the focus for TM mode is around $\lambda/2$ from the backside of the

[4]The authors would like to thanks to Borislav Vasić, Institute of Physics Belgrade, Serbia, for help in simulations.

photonic crystal, while for TE mode it is exactly on the back side. The length of a jet for TM mode is about 2 times more than for TE mode. This is simple result could be interesting, especially since it is not easy to manufacture birenfringent focusing lens. A specific feature of the birefringent photonic jets is a possibility to form *two separate jets* for the TE- and TM-polarized waves.

References

1. Minin, O. V., & Minin, I. V. (2004). *Diffractional optics of millimeter waves*. Bristol: Institute of Physics Publishing.
2. Minin, I. V., Minin, O. V., Gagnon, N., & Petosa, A. (2007). Investigation of the resolution of phase correcting Fresnel lenses with small focal length-to-diameter ratio and subwavelength focus. In *Proceeding of the EMTS 2007*, Canada, Ottawa (URSI), July 26–28, 2007.
3. Joannopoulos, J. D., Meade, R. D., & Winn, J. N. (1995). *Photonic crystals: molding the flow of light*. Princeton: Princeton University Press.
4. Sakoda, K. (2005). *Optical properties of photonic crystals* (2nd ed.). New York: Springer.
5. Antonakakis, T., Craster, R. V., & Guenneau, S. (2013). Asymptotics for metamaterials and photonic crystals. *Proceedings of the Royal Society of London A, 469*, 20120533.
6. Goss Levi, B. (1999). Progress made in near-field imaging with light from a sharp tip. *Physics Today*, *52*, 18.
7. Flück, E., van Hulst, N. F., Vos, W. L., & Kuipers, L. (2003). Near-field optical investigation of three-dimensional photonic crystals. *Physical Review E, 68*, 015601.
8. Li, C., Holt, M., & Efros, A. L. (2006). Far-field imagimg by the Veselago lens made of a photonic crystal. *Journal of the Optical Society B, 23*(3), 490–497.
9. Daschner, F., Knöchel, R., Foca, E., Carstensen, J., Sergentu, V. V., Föll, H., & Tiginyanu, I. M. (2006). Photonic crystals as host material for a new generation of microwave components. *Advances in Radio Science, 4*, 17–19.
10. Minin, I. V., Minin, O. V., Triandaphilov, Y. R., & Kotlyar, V. V. (2008). Subwavelength diffractive photonic crystal lens. *Progress in Electromagnetics Research B (PIER B)*, *7*, 257–264.
11. Minin, I. V., Minin, O. V., Triandaphilov, Y. R., & Kotlyar, V. V. (2008). Focusing properties of two types of diffractive photonic crystal lens. *Optical Memory & Neural Networks (Information Optics), 17*(3), 244–248.
12. Minin, I. V., Minin, O. V., Triandaphilov, Y. R., & Kotlyar, V. V. (2008). Subwavelength diffractive photonic crystal lens. In *Proceedings of the 2008 China-Japan Joint Microwave Conference* (pp. 756–757). Shanghai, China, September 10–12, 2008.
13. Minin, I. V., Minin, O. V., Triandaphilov, Y. R., & Kotlyar, V. V. (2008). Subwavelength Diffractive photonic crystal lens. In *Proceedings of the International Conference on Mathematical Physics and Its Applications—Steklov Mathematical Institute of the Russian Academy of Sciences* (p. 128). Samara State University. Samara, Russia, September 8–13, 2008.
14. Yee, K. S. (1966). Numerical solution of initial boundary value problems involving Maxwell's equations in isotropic media. *IEEE Transactions on Antennas and Propagation, AP-14*, 302–307.
15. Umashankar, K. R., & Taflove, A. (1982). A novel method to analyze electromagnetic scattering of complex objects. *IEEE Transactions on Electromagnetic Compatibility, 24*(4), 397–405.
16. Berenger, J. P. (1994). A perfectly matched layer for the absorption of electromagnetic waves. *Computational Physics, 114*, 185–200.
17. Triandafilov, Y. R., & Kotlyar, V. V. (2007). Photonic-crystal Mikaelian lens. *Computer Optics*, *31*(3), 27–31.

18. Luo, C., Johnson, S. G., Joannopoulos, J. D., & Pendry, J. B. (2002). All-angle negative refraction without negative effective index. *Physical Review B, 65*, 201104(R).
19. Fang, Y., & Shen, T. (2007). Diverse imaging of photonic crystal be the effects of channeling and partial band gap. *Optik, 118*, 100–102.
20. Fang, Y.-T., & Sun, H.-J. (2005). Imaging by photonic crystal using reflection and negative refraction. *Chinese Physics Letters, 22*(10), 2674–2676.
21. Li, Z.-Y., & Lin, L.-L. (2003). Evaluation of lensing in photonic crystal slabs exhibiting negative refraction. *Physical Review B, 68*, 245110.
22. Luo, C., Johnson, S. G., Joannopoulos, J. D., & Pendry, J. B. (2003). Subwavelength imaging in photonic crystals. *Physical Review B, 68*, 045115.
23. Kiasat, Y., Szabo, Zs., & Li, E. P. (2013). Sub-wavelength imaging with non-spherical plasmonic nano-particles. In *Proceedings Of the 4th International Conference on Metamaterials, Photonic Crystals and Plasmonics* (pp. 49–50). Sharjah, United Arab Emirates, March 18–22, 2013.
24. Meriakri, V. V. & Nikitin, I. P. (1989). Man-made dielectric with dispersion in the millimeter wavelength band. Quasioptical devices in millimeter and sub-millimeter wavelength ranges. In *Proceedings of IRE* (pp. 65–70). Har'kov.
25. Petosa, A., Ittipiboon, A., & Thirakoune, S. (2002). Perforated dielectric resonator antennas. *Electronics Letters, 38*(24), 1493–1495.
26. Zhang, Y., & Kishk, A. A. (2007). Analysis of dielectric resonator antenna arrays with supporting perforated rods. In *2nd European Conference on Antennas and Propagation, EuCAP 2007* (pp. 1–5).
27. Fabre, N., Fasquel, S., Legrand, C., Melique, X., Muller, M., Francois, M., et al. (2006). Towards focusing using photonic crystal flat lens. *Opto-Electronics Review, 14*(3), 225–232.
28. Pacheco-Peña, V., Beruete, M., Minin, I. V., & Minin, O. V. (2014). Terajets produced by 3D dielectric cuboids. *Applied Physics Letters, 105*, 084102.
29. Joannopoulos, J., Meade, R., & Winn, J. (1995). *Photonic crystals.* Princeton: Princeton University Press.
30. Lourtioz, J.-M., Benisty, H., Berger, V., Gerard, J.-M., Maystre, D., & Tchelnokov, A. (2008). *Photonic crystals. Towards nanoscale photonic devices*. Berlin: Springer.
31. Ehrhardt, M. (Ed.), *Wave propagation in periodic media. Analysis, numerical techniques and practical applications*. Sharjah: Bentham Science Publishers.
32. Gong, Q., & Hu, X. (2014). *Photonic crystals: principles and applications* (366 p). New York: CRC Press.

Chapter 4
Photonic Jets Formation by Non Spherical Axially and Spatially Asymmetric 3D Dielectric Particles

Abstract In this Chapter we address a fundamental question: is a spherical-aided shape of dielectric particle to form a photonic jet unique or the spherical shape of particle may be extended to other form? With the aim to obtain subwavelength focusing, an alternative mechanism (in contrast to spherical-aided particles) to produce photonic jets by using 3D and 2D dielectric cuboids are discussed. The principle possibility of generation and management of photonic jets parameters (including 3D) by choosing the particle of 3D arbitrary shape free of axial spatial symmetry are shown for the first time. Also for the first time the possibility of photonic jet formation in the interaction of a plane wave front with a particle located on a reflecting substrate in the "reflection" mode (flat focusing mirror) are offered.

Keywords Photonic jet · 3D particle · Reflection mode · Cuboid · Particle shape · Jet parameters

Introduction

As it was mentioned the fundamental Rayleigh criterion for optical systems resolution implies that the minimal size of the observable object is somewhat smaller than the wavelength of the applied radiation being basically limited by its diffraction [1]. Some groups of researchers around the World are now making an attempt to overcome the diffraction limit, i.e. to focus radiation in a spot smaller than Airy disk [1]. To overcome the diffraction limit some solutions were offered.

Visible and IR radiation interaction with transparent media has been well studied and known since a long time ago [2]. However, quite recently for the first time (in work [3]) 'photon nanojet'[1] (PNJ) effect was observed in process of investigating laser radiation scattering on transparent quartz microcylinders [4] and, later, on spherical particles. *Dielectric particles which form a photonic jet* play a key role as components able to provide both the high density of the electromagnetic power and the integration with receivers and signal-processing circuits.

[1]It is interesting to note that the term "nanojet" now is used, for example, also in the problem of laser pulse impact to locally melt surface. See: Valev et al. [70].

I. Minin and O. Minin, *Diffractive Optics and Nanophotonics*,
SpringerBriefs in Physics, DOI 10.1007/978-3-319-24253-8_4

Physics of Photonic Jet Formation in Spherical-Based Dielectric Particle

Consider in brief the structure of dielectric spherical particle PNJ. When radiation interacts with a dielectric sphere, the radiation is scattered in the far field. However in the near field, the same radiation can be highly concentrated. In the optimum case, the radiated power density can be concentrated more than 200 times in a propagative beam, having a low divergence and a full width at half-maximum that is smaller than the wavelength. This beam is known as a "photonic jet". Its physical laws are different to those of classical geometrical optics.

According to Mie theory, the optical field both inside and outside the low absorbing sphere, subject to the light wave is characterized by focusing spaces called inner and outer focuses of the field.[2] They result from the spherical particle surface curvature, which causes relevant deformations of the incident phase wave front. Spherical microparticle, thus, acts as a refractive microlens, focusing light within subwavelength volume [5].

From the geometrical-optics approximations it is followed if $n > 2$ the focus is inside sphere and if $1 < n < 2$ the focus is outside of sphere, where n—refractive index of sphere (Fig. 4.1). In the physical optics approximation the choice of the refractive index contrast (RIC) depends on the radius of the sphere in the unit of the wavelength (usually it is found to be about 2). So when the focus point is just on the boundary of the sphere a so called phenomenon of "photonic jet" occurs—on the free space shadow side of the surface, an intense optical jet-like region is generated.

It could be noted that for mesoscale or nanoscale objects, classical geometrical optics fails for IR and visible light because the interactions of such objects with light waves are described by near-field optics [6].

Based on analyzing of the numerical aperture and the spherical aberration of the microsphere with radius more than wavelength the authors [7] demonstrate that the microsphere has negligible spherical aberration and high *NA* when *RIC* between the microsphere and its surrounding medium is about from 1.5 to 1.75. The reason is due to the spherical aberration compensation arising from the positive spherical aberration caused by the surface shape of the microsphere and the *RIC* and the negative spherical aberration caused by the focal shifts due to the wavelength scale dimension of the microsphere. It also was shown that, only within the approximate region of $RIC \in [1.5\text{–}1.75]$ with the proper diameter d of microsphere ($d > 2.25\lambda$), the microsphere[3] generate a near-field focal spot with lateral resolution slightly beyond $\lambda/2n$, which is also the lateral resolution limit of the dielectric microsphere.

[2]Chang and Pan [71].

[3]It is interesting to note that the optimization of spherical particle shape runs towards a mixture with a Fresnel-type lens, described in Chap. 1. See: Paganini et al. [72]

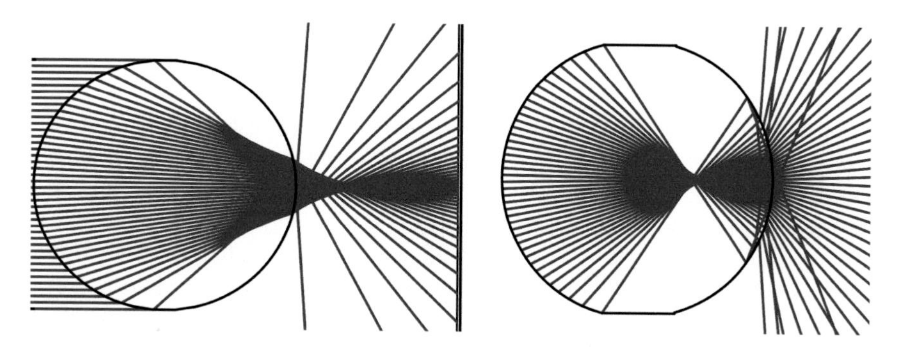

Fig. 4.1 Ray-trasing of sphere focusing: RIC = 1,5 (*left*) and RIC = 3 (*right*)

But the method used in [7] is invalid for particle with diameter less than 2 wavelengths. Because of the scale of the problem, this focus point does not obey to the geometrical laws for dielectric spherical particles with a diameter size that can be compared with the wavelength. Thus these effects can not be predicted on the basis of geometrical optics or scalar diffraction theory, and it is essential to study the propagation of electromagnetic waves through such elements using the Maxwell equations. Photonic jet arises on the shadow surface of dielectric microspheric particles in so called near-field diffraction region. It features strong spatial localization and high-intensity of optical field in the focusing area[4] with spatial resolution up to the third of the wavelength that is lower than the classical diffraction limit [5]. It should be understood that despite of its name—the photonic jet—this wave structure is not related to the quantum nature of radiation.

Basic parameters to optimize spheroidal particles PNJ characteristics embrace the incident wave front shape, Mie-parameters, and RIC [7–10] (the length of the photonic jet depends strongly on the RIC).

The possibility of producing photonic nanojets was studied as regards dielectric elliptical nanoparticles [12], multilayer heterogeneous microspherical particles with radial refractive index gradient [13, 14], and "truncated" elipsoids [15]. Basic characteristics of PNJ formed in the neighborhood of heterogeneous dielectric microspheres and microcylinders subject to laser radiation were studied, particularly, in work [16]. The first experiments on the direct observation of photonic jets were conducted in microwave range [17, 18]. It was also shown that the photonic nanojet effect is observed in case of interaction of surface plasmon waves with plane cylinder [19]. A hybrid hemispherical based linear-microdisk shape and a

[4]A breakthrough in far-field sub-wavelength imaging with white light source was reported in Wang et al. [50]. The idea was based on overcoming the diffraction limit by using the ordinary SiO_2 microspheres as superlenses which forms a photonic nanojet. Later [73] it is shown that only evanescent waves, which carry the high frequency spatial sub-wavelength information, are responsible for the formation of near field image. It is also demonstrated analytically that while the evanescent waves improve the resolution of the real image, the remarkable imaging performance is due to nanoscope's sub-wavelength near field focusing size.

hybrid hemispherical based microdisk-parabolic shape particles for photonic jet formation are considered at [20]. The attempt to classify the photonic jets in appearance was made in [21].

But *it has been assumed that such microparticles should have high degree of spatial axial symmetry of the shape (spheres, spheroids, cylinders, disks).* Also it should be noted that all particles shape which were investigated in the literature was described by a second-order algebraic surface [11]. Thus the bulk of theoretical studies devoted to nanojets has been predominantly focused on the case of dielectric particles as hemi-spherical aided shape. In this chapter we intend to fill the gap.

Although analytical solutions of the vector diffraction problem can be obtained for selected objects (sphere, halfplane, cylinder) [1, 2] the boundary conditions on the electromagnetic field for other dielectric structures makes the analytic solution impossible. In order to evaluate the focusing performance of the structure, the transient solver of the commercial software CST Microwave Studio™ was used along with an extra fine hexahedral mesh with a mínimum mesh size of $\lambda_0/45$ [22–24].

Cuboid Dielectric Particle

The fundamental question in application to photonic jet is: is a spherical-aided shape of dielectric particle to form a photonic jet unique or the spherical shape of particle may be extended to other form?

To manage the whole set of PNJ parameters and optimize their characteristics additional free parameters are needed. In particular, PNJ parameters management by, for example, choosing the particle shape (cube, triangle, pyramid, hexagonal, axicon, etc.) has not been studied yet.

In work [25] it was shown for the first time that photonic terajets (analog of optical PNJ) may be formed also in case of plane wave front interaction with mesoscale cubic dielectrical structure. The 3D dielectric cuboid is schematically presented in Fig. 4.2 and has lateral dimensions $L = \lambda_0$, along x and y axes while the dimension H is selected to be $(1\ldots1.3)\lambda_0$. In the simulation the 3D cuboid is illuminated with a vertically (E_y) polarized plane wave (see Fig. 4.2) at 0.1 THz (λ_0 = 3 mm). Moreover, vacuum (n_0 = 1) is used as the background medium and expanded open boundary conditions are used in order to insert the 3D cuboid within an infinite medium.

Based on this, the performance of the terajets is evaluated using homogeneous 3D cuboids with different values of refractive index (n). Numerical results of the power distribution on the yz-plane/E-plane (left column) and xz-plane/H-plane (right column) are shown in Fig. 4.2. Similar to photonic nanojets produced by cylindrical and spherical dielectrics [11, 14], it can be observed that the terajet is located inside the dielectric structure for higher values of refractive index.

The best resolution was obtained for n = 1.75 with $FWHM_{x,y} < 0.4\lambda_0$ and an intensity enhancement of ~15 times the incident plane wave at the output surface

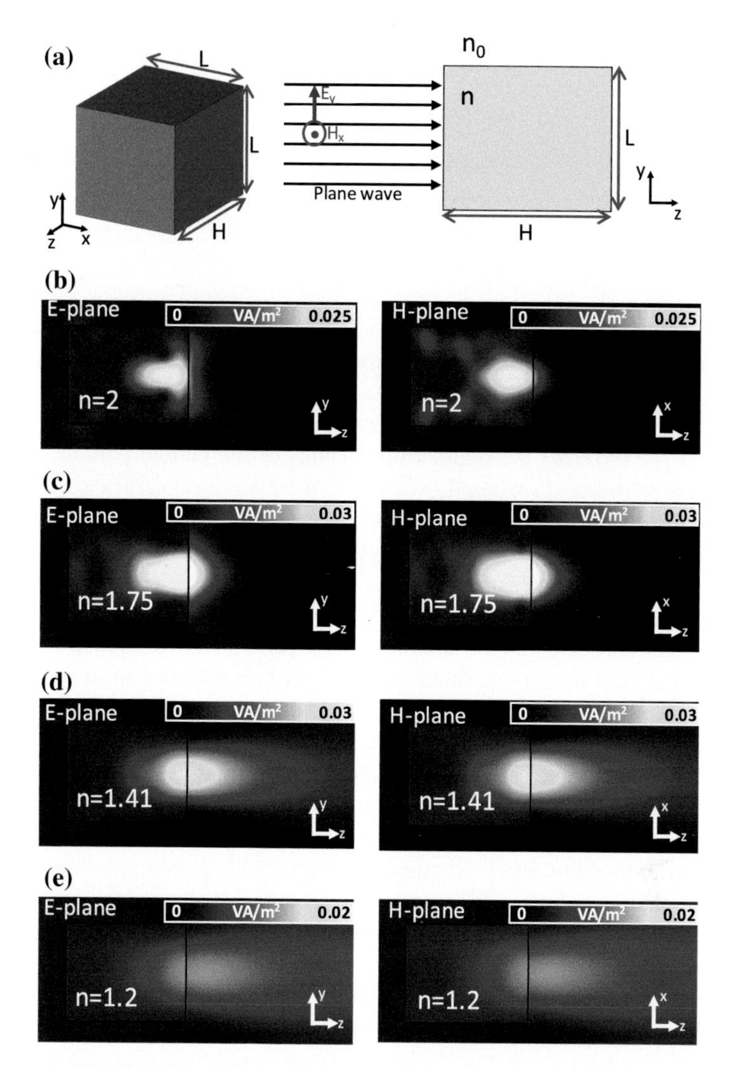

Fig. 4.2 Schematic representation of the proposed 3D dielectric cuboid with dimensions $L = \lambda_0$ and $H = 1.2\lambda_0$: (**a**) Perspective and (**b**) lateral view. Numerical simulations of the terajet performance for different values of the refractive index: (**b**) $n = 2$, (**c**) $n = 1.75$, (**d**) $n = 1.41$ and (**e**) $n = 1.2$ on the *yz*-plane/*E*-plane (*left column*) and *xz*-plane/*H*-plane (*right column*)

of the 3D cuboid. However, for $n = 1.41$ a quasi-symmetric terajet was obtained and an intensity of ~ 10 times the power of the incident plane wave. We can define the parameter *terajet exploration range*, denoted as Δz, as the distance from the surface at which the intensity enhancement has decayed to half its maximum value (note that we are implicitly assuming here that the maximum appears at the surface, since this is the preferred scenario for microscopy applications). From the results it is followed [25] that the exploration range for $n = 1.41$ is $\Delta z = 0.72\lambda_0$ notably larger

Table 4.1 Parameters of PNJ versus coboid dimensions

H/λ	Min FWHM, λ	W/H
0.62	0.405	1.3
0.8	0.395	1.17
1.0	0.385	1.0
1.3	0.41	0.81
1.5	0.43	0.72

than for $n = 1.75$, $\Delta z = 0.16\ \lambda_0$. It could be noted that terajet have the same electric field polarisation as the incident wavefront. Thus, in order to generate jets at the surface of the dielectric cuboid RIC less than 2:1 is required, in agreement with the results of photonic nanojets using circular 2D/3D dielectrics [3, 4].

Additional simulations shown that from the point of view of minimal photonic jet diameter (i.e.FWHM) the optimal dimensions of the cuboid particle is a cubic with dimension of LxLxL, where L—is a free space wavelength. The correspondent simulation results for the cubic with refractive index contrast n = 1,46 are shown in the Table 4.1 below (where: H—the height and W—the width of cuboid sides).

It can be seen that increasing height of cuboid optimal value of W/H decreases and minimal FWHM is obtained for W/H = 1 and equal to 0,385 wavelength.

The comparison of common simulated jet's parameters for sphere and cuboid has shown that there are 2 main differences: (1) the cuboid-aided focal spots are less elliptical than the sphere-aided focal spots; (2) with increasing sphere diameter and cuboid's side, the photonic jet's length measured as the intensity half-maximum behaves differently: increasing for the cuboid and decreasing for the sphere.

Physical principle of PNJ formation in this case consists in the following: when falling on the cubic particle the plane wave penetrates into the dielectrical material. As near the cuboid edge, inner radiation propagates with greater phase velocity than in its centre, the phase incursion arising between different areas of the incident wave causes deformation of radiation wave front. At certain parameters of the cubic particle the wave front acquires positive curvature (radiation propagates from the cuboid edge to its centre), that conforms to the condition of the radiation focusing. Thus, as opposed to spheroids, it is the incident wave diffraction, not the refraction, that plays the main role in terajet formation on such structures. So to the PNJ formation the diffraction effects in the scattering of radiation are influence. First of all—the interference of passing through the particle and refracted waves are important.

Backscattering Enhancement Evaluation

Similarly to nanojets produced by cylindrical and spherical dielectrics [30, 31], we evaluate the backscattering enhancement when a metal particle is introduced within the terajet with respect to the backscattering without the metal particle [26, 27]. The backscattering enhancement of 2D and 3D dielectric cuboids were compared.

Note that by using one dimension larger than the rest of the 2D cuboid, the focusing performance along the *yz*-plane/*E*-plane is changed and a cylindrical Terajet ("teraknife") is obtained [26]. It was shown that higher values are obtained for the 3D case, in good agreement with previous results [3, 4, 18]. However, even though lower values of backscattering enhancement are achieved using 2D cuboids, these structures are able to detect particles placed at different positions along the cylindrical teraknife by recording its backscattering intensity [36].

The existence of terajets as well as the backscattering enhancement was also experimentally demonstrated at sub-THz frequencies. As it has been explained previously, the 3D cuboid here proposed can be scaled from microwaves up to optical frequencies because of all dimensions are given in the wavelength units. The measurements of the backscattering intensity were carried out using the method of movable probe [28, 29]. It was shown that both simulation and experimental results were in good agreement (with a maximum error between them of 7 % and a backscattering enhancement of $\sim$0.57 dB).

Thus, 3D and 3D dielectric cuboids working at terahertz frequencies with the ability to generate terajets at its output surface under plane wave illumination have been studied.[5] The structure here proposed can be used at microwaves and also as a scaled model at optical frequencies, simplifying the fabrication process compared with spherical dielectrics and could find applications in novel microscopy devices.

Multifrequency Focusing and Wide Angular Scanning of Terajets

It was also demonstrated [32] that terajets formation on the basis of dielectrical cuboids is possible not only on fundamental harmonic, but also on other even frequency harmonics and also in case of the plane wave front oblique incidence. Dielectric cuboid here acts as a flat lens with PNJ being a focus.

One important aspect to evaluate is the frequency response of the terajet keeping the dimensions constant. Simulation results are shown in Fig. 4.3. By comparing both *E* and *H* planes for each frequency in Fig. 4.3, an interesting feature can be observed: a quasi-symmetric focus is maintained in all cases, demonstrating that it is possible to use the same structure at different harmonics while maintaining its resolution.

The highest intensity was obtained for the first harmonic. The terajet generated at the second harmonic is slightly deteriorated and a clear secondary peak is observed away from the output surface of the cuboid at $z \sim 2.1\lambda_2$' (= $0.7\lambda_0$'), while the focusing properties of the terajet are not changed for the second harmonic.

[5] Interestingly, that the similar jets from cuboid based dielectric objects, but fabricated of silica with refractive index $n = 1.46$ on a substrate were obtained at optical frequencie: Measurement of photonic nanojet generated by square-profile microstep // *Proc. SPIE* 9448: Optical Technologies in Biophysics and Medicine XVI; Laser Physics and Photonics XVI; and Computational Biophysics, 94482C (March 19, 2015).

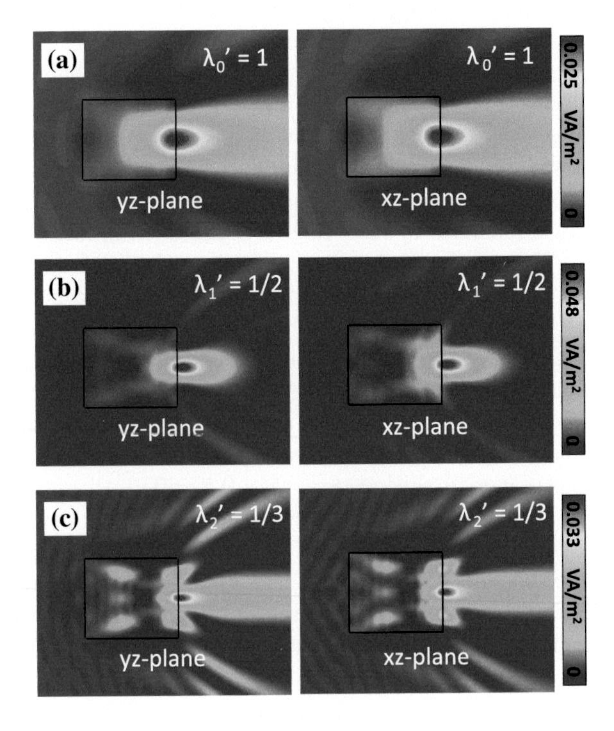

Fig. 4.3 Simulation results of the power distribution in the *E*-plane (*left column*) and *H*-plane (*right column*) when the 3D dielectric cuboid with the initial dimensions (*L* along both *x*- and *y*- axes and *H* along z-axis) is illuminated with a planewave under normal incidence at: (**a**) the fundamental frequency $\lambda_0' = 1$ and the (**b**) first $\lambda_1' = 1/2$ and (**c**) second $\lambda_2' = 1/3$ frequency harmonics

Table 4.2 Simulation results of the terajet performance using the 3D dielectric cuboid with refractive index n = 1.41 using planewave illumination under normal incidence

Frequency	FL	Enhancement	$FWHM_x$	Δ_z
$f_0' = 1$ ($\lambda_0' = 1$)	$-0.077\lambda_0'$	~10	$0.47\lambda_0'$	$0.72\lambda_0'$
$f_1' = 2$ ($\lambda_1' = 1/2$)	$0.16\lambda_1'$	~18	$0.52\lambda_1'$	$0.86\lambda_1'$
$f_2' = 3$ ($\lambda_2' = 1/3$)	$-0.067\lambda_2'$	~14	$0.62\lambda_2'$	$0.58\lambda_2'$

The power enhancement (calculated as the received power at the focal position compared with the power received without the 3D dielectric cuboid) is ~10, ~18 and ~14 times the incident planewave at each frequency. Nevertheless, sub-wavelength resolution is achievable with all the frequencies here evaluated. A summary of these results is shown in Table 4.2, where: FL is the focal length, $FWHM_x$ is the full width at half maximum along the *x*-axis just at the output surface of the 3D dielectric cuboid for each harmonic, Δ_z is the exploration range.

Let's now consider the focusing possibilities of dielectric cuboids with fixed wavelength but different dimensions (it is equivalent to the frequency harmonic properties). The results of FDTD simulations are shown in the Fig. 4.4. The main focusing characteristics of photonic jets are given in the Table 4.3.

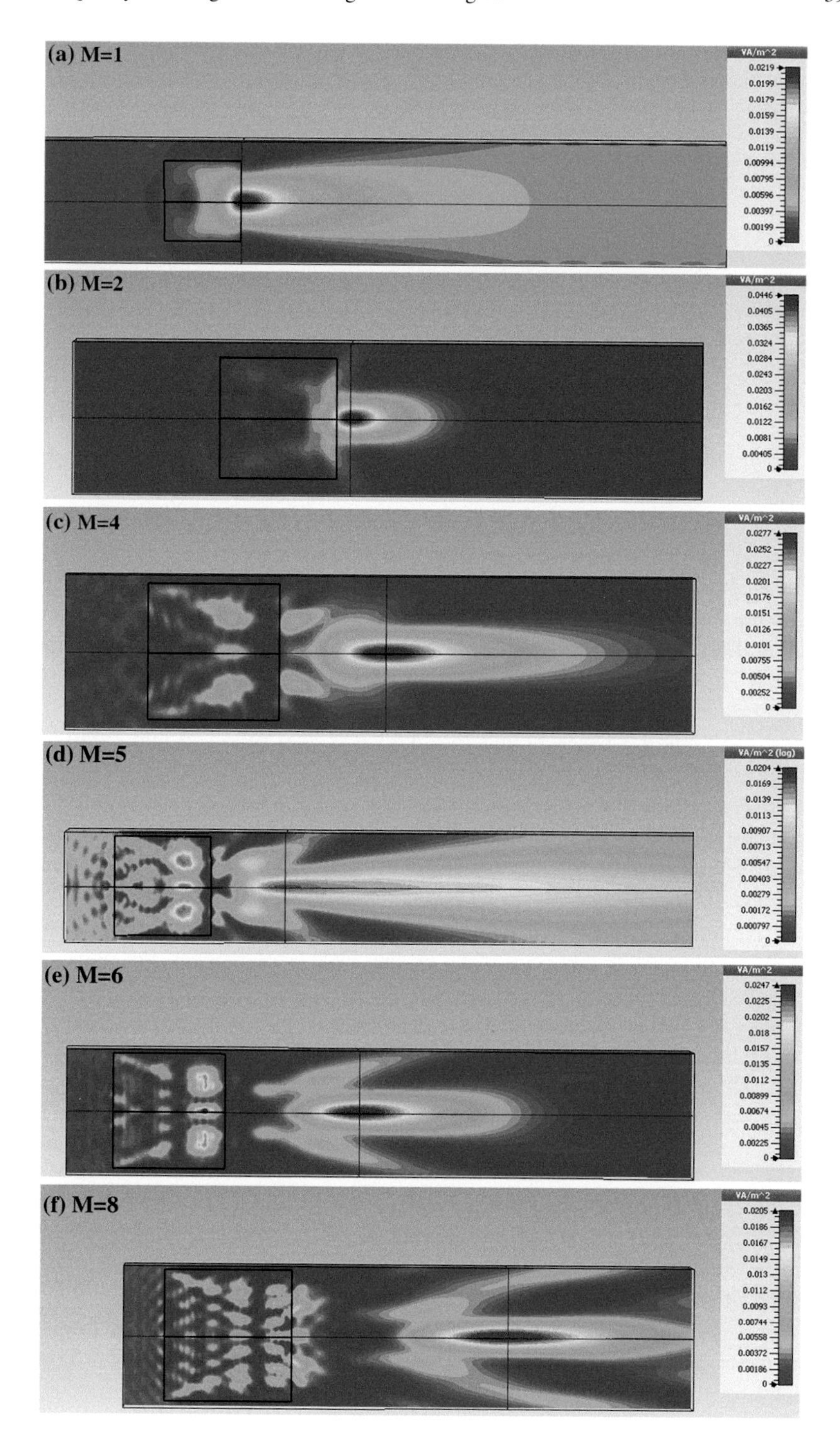
(a) M=1
VA/m^2
0.0219
0.0199
0.0179
0.0159
0.0139
0.0119
0.00994
0.00795
0.00596
0.00397
0.00199
0
(b) M=2
VA/m^2
0.0446
0.0405
0.0365
0.0324
0.0284
0.0243
0.0203
0.0162
0.0122
0.0081
0.00405
0
(c) M=4
VA/m^2
0.0277
0.0252
0.0227
0.0201
0.0176
0.0151
0.0126
0.0101
0.00755
0.00504
0.00252
0
(d) M=5
VA/m^2 (log)
0.0204
0.0169
0.0139
0.0113
0.00907
0.00713
0.00547
0.00403
0.00279
0.00172
0.000797
0
(e) M=6
VA/m^2
0.0247
0.0225
0.0202
0.018
0.0157
0.0135
0.0112
0.00899
0.00674
0.0045
0.00225
0
(f) M=8
VA/m^2
0.0205
0.0186
0.0167
0.0149
0.013
0.0112
0.0093
0.00744
0.00558
0.00372
0.00186
0

◀ **Fig. 4.4** Field intensity distributions in ZX plane for dielectric cuboids with different dimensions: the initial dimension of the cuboid was 1 × 1 × 1 in wavelength and the next cuboids dimensions were multiply by factor which shown near the correspondent figure. Factor M = 1 corresponds to the original dimensions of the 3D cuboid. The factors M = 2–8 correspond to simulation results when the dimensions of the original cuboid are multiplied for this value. **a** M = 1, **b** M = 2, **c** M = 4, **d** M = 5, **e** M = 6, **f** M = 8

Table 4.3 Focusing characteristics of photonic jets

Cuboid dimensions, multiply factor M	FWHM x, λ	FWHM y, λ	FWHM z, λ
1	0.45	0.43	1.1
2	0.51	0.49	1.2
4	1	0.95	4.74
5	*1.06*	*1.45*	>22
6	1.4	1.16	7.83
8	1.77	1.56	13

The analysis of the results shown that the focusing properties are saved when the size of the cube increases by an even number of times, and the localization of radiation deteriorates with increasing the size of cube in an odd number of times. Also it is interesting to note that ellipticity of the photonic jet (resolution x/y) is almost constant up to multiply factor of M = 4.

From the table 4.3 (M = 5) and Fig. 4.4d it is also followed that "ultra-long" photonic jet is form. This effect is due to two effects: first is a photonic jet formation from dielectric cube and second is a scattering on cube boundary in near field.

It is also followed from the simulations that changing the dimension of the dielectric cuboid (factor M), the focus point is moved from inside to outside the cuboid and the focal length increases as dimension increases. At large dimension of M, the photonic jet is formed rather far from the shadow of the cuboid, and a decrease in dimension the coordinate of the focal spot reaches the cube edge. The results show that the length of photonic jet is elongated greatly with high value of M. As for FWHM vs factor M these dependences may be approximated by the following linear expressions (as it follows from the Table 4.3):

$$\mathrm{X} = 0,434\mathrm{M} + 0,413, \quad \mathrm{Y} = 0,382\mathrm{M} + 0,583,$$

where X, Y—the FWHM along X, Y axis in the unit of FWHM along X at M = 1, respectively. The approximation of $FWHM_{x,y} \approx \alpha_{x,y}M$ for M = even is also valid. So the location and 3D size of the photonic jet depend on the dimension of the dielectric cuboids'. Thus the properties of photonic jet can be controlled by the variation of the dimensions of the dielectric cuboid.

The results described above showed a rich structure in the high spatial frequency components of the photonic jet. As it can see in direct space, each propagative spatial frequency corresponds to propagation at given angle with respect to the beam axis. An example of intensity map of a photonic jet is presented on a colored

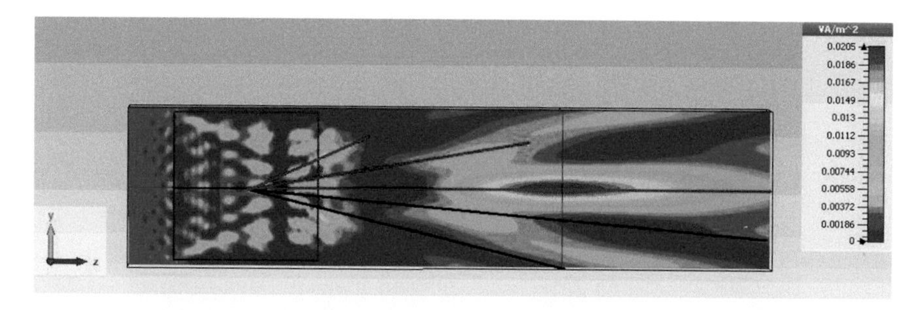

Fig. 4.5 Scattered intensity of a photonic jet produced by a dielectric cuboid with factor M = 8. The angles corresponding to the maxima and minima in the spatial frequency are displayed in direct space, respectively, by *red* and *black lines*

map in Fig. 4.5. The angles corresponding respectively to the first few maxima and the minima of the field intensity are displayed respectively with red lines and black lines in this figure. The maxima correspond to high intensity angles while the minima correspond to angles of low intensity regions. The maxima in the spectral distribution can therefore be associated with the presence of secondary lobes in the direct field structure. The first secondary lobes tend to confine the central lobe into a low divergent beam, while the secondary lobes with high transverse components tend to reduce the length and the waist of photonic jets.

Since all the results shown before have been obtained considering normal illumination, it is important to evaluate the focusing properties of the under oblique incidence [32]. Simulation results of the normalized power distribution in the $xz(H)$ plane for input angles from 0° to 80° are shown in Fig. 4.6 particularized only for the fundamental frequency. It can be observed that the terajet is deflected when the input angle is changed, as expected.

It was also shown that the beam deviation factor (output angle/input angle) has an almost constant slope with a value of 1 for input angles from 0° to 45°; moreover, the intensity of the terajet at the output face of the 3D cuboid is not reduced for these angles, contrarily to the behavior at 60° and 80°. For the case of the first and second frequency harmonics the same dependences were obtained, demonstrating that the terajet produced by the cuboid has a robust performance even under oblique incidence with relatively high incidence angles.

Experimental and simulation results have a good agreement for the fundamental and first frequency harmonic, respectively. Thus, the capability to generate multi-frequency terajets using 3D dielectric cuboids working at frequency harmonics were demonstrated at THz frequencies as a scale model for optics.

Breaking the imaging symmetry in cuboid-aided photonic jet. It is well known that conventional lenses have symmetrical imaging properties along forward and backward directions. In this content cuboid performs as either converging lenses or diverging lenses depending on the illumination directions. These new imaging properties, including symmetry breaking as well as the super resolving power, significantly expand the horizon of imaging optics and optical system design.

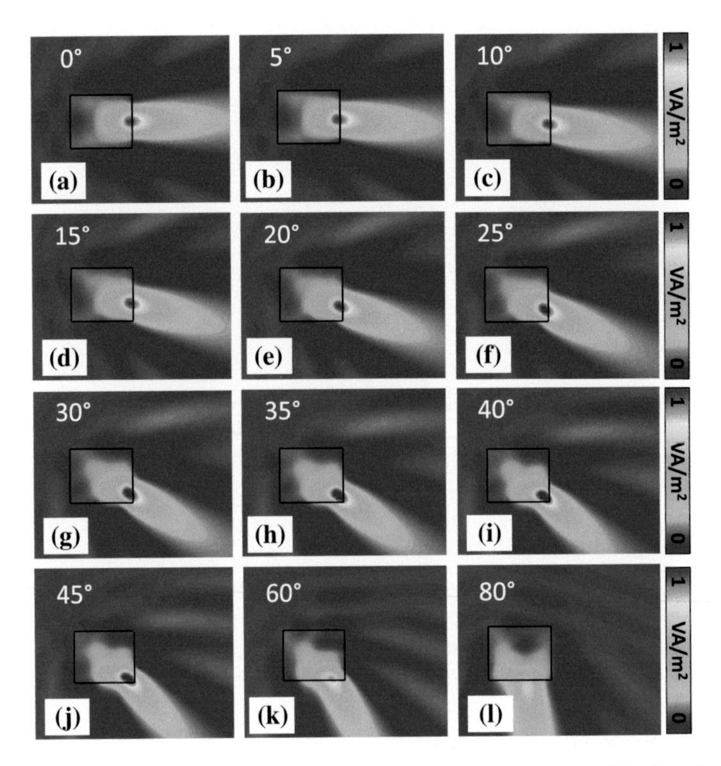

Fig. 4.6 Simulation results of the normalized power distribution in the *xz*(*H*) plane for the 3D dielectric cuboid under oblique incidence at the fundamental frequency (λ‘ = 1): (**a**) 0°, (**b**) 5°, (**c**) 10°, (**d**) 15°, (**e**) 20°, (**f**) 25°, (**g**) 30°, (**h**) 35°, (**i**) 40°, (**j**) 45°, (**k**) 60° and (**l**) 80°. Simulation results are normalized with respect to the maximum power obtained under normal illumination (0°)

Polarization Properties of Mesoscale Regular Hexahedron-Aided Terajet

Polarization of incident wavefront is an important parameter, which is helpful in the search for longitudinally and transversally subwavelength photonic jets.

Beam shaping of nanojet by polarization engineering were considered in [33], where the authors used a 2-μ-diameter latex sphere (D/λ = 3) with RIC = 1,2. It has been shown that when the microsphere is illuminated by linear and circular polarization beams, the axial field intensity profile is the same. Azimuthal polarization incident beam induces a doughnut beam along the optical axis and compared with linear and circular polarizations, both the transverse FWHM and axial half-decay length of the photonic nanojet are clearly decreased by the radial polarization incident beam, with the maximum intensity being close to the microsphere.

The changes of the photonic jet demonstrate its maximum field intensity and quality criterion of the jet (Q). The complex characteristic of a PNJ can be given

Table 4.4 PNJ versus polarization of illumination wavefront

Polarization/FWHM	X, λ	Y, λ	Z, λ	Elipticity	Q
Linear	0.46	0.43	1.08	1.07	2.51
Circular	0.51	0.52	1.08	0.98	3.83
Diagonal	0.44	0.44	1.08	1.0	4.89

with the aid of the modified so-called "quality criterion Q" [34], which combines all relevant jet parameters. We define Q as: $Q = L_{jet} I_{\max} / \min(FWHM_{x,y})$— the PNJ beam length L_{jet} is FWHM along z-axis, I—maximal value of field intensity along the photonic jet (in Table 4.4—relative to cuboid with linear polarization). Thus the photonic jet's length was calculated as the $FWHM_z$ of intensity outside the particle (i.e., if the maximum intensity was found inside the particle, the half-maximum was counted from the surface).

Simulations showed that for circular polarization the electric field maximum is moved away from the surface of the 3D cuboid (it is a regular hexahedron of side $L = \lambda_O$, with refractive index contrast $n = 1.46$) along the optical axis (z). The parameters of PNJ for different polarization state of incident wavefront for the cuboid are shown in the Table 4.4.

For the circular polarization of incident wavefront the localized field intensity (photonic jet) have no the subdiffractive dimensions instead of linear polarization.

This physical phenomenon for cuboids briefly described above could be a significant implement in the fields of photonic circuit. The photonic jets, for example, permit the lightwave coupling from the photonic molecule into other photonic components such as planar waveguides or coaxial cables. The cuboid-aided photonic jet allows also increasing the resolution of micro- solid lens in optical microscopy,[6] etc.

The Possibilities of Curved Photonic Jet Formation (Photonic Hook)

To the best of our knowledge, no research dealing with the generation of curved terajets as a near-field curved optical flux (PNJ) by means of mesoscale dielectric particles has been published so far.

In optics the propagating along the parabolic path, the Airy beams were used in the areas of micromanipulation (an "optical spade") [35] and generation of curved plasma channels [36]. Let's briefly consider the dielectric cuboid (refractive index contrast of 1.46) with dimensions of 3 × 3 × 3 wavelength with triangle bar with dimensions of 1 × 3 × 3.16 wavelength. It corresponds to oblique illumination under the 18.435°.

---

[6]Hou et al. [74].

The radiation passing through the transparent dielectric object obtains a phase delay compared with the radiation passing near such object. Therefore, the wave front is concave and convergent. So the focusing condition of the wavefront is observed. But since this is not a spherical wave front, but the aberration, the focus area extends along the Z axis.

From the Fig. 4.7 it is followed that in the propagation of the focal spot is not only shifted from the optical axis, but changes its shape, similar to a light beam with astigmatism. The parameters of the jets are shown in the Table 4.5.

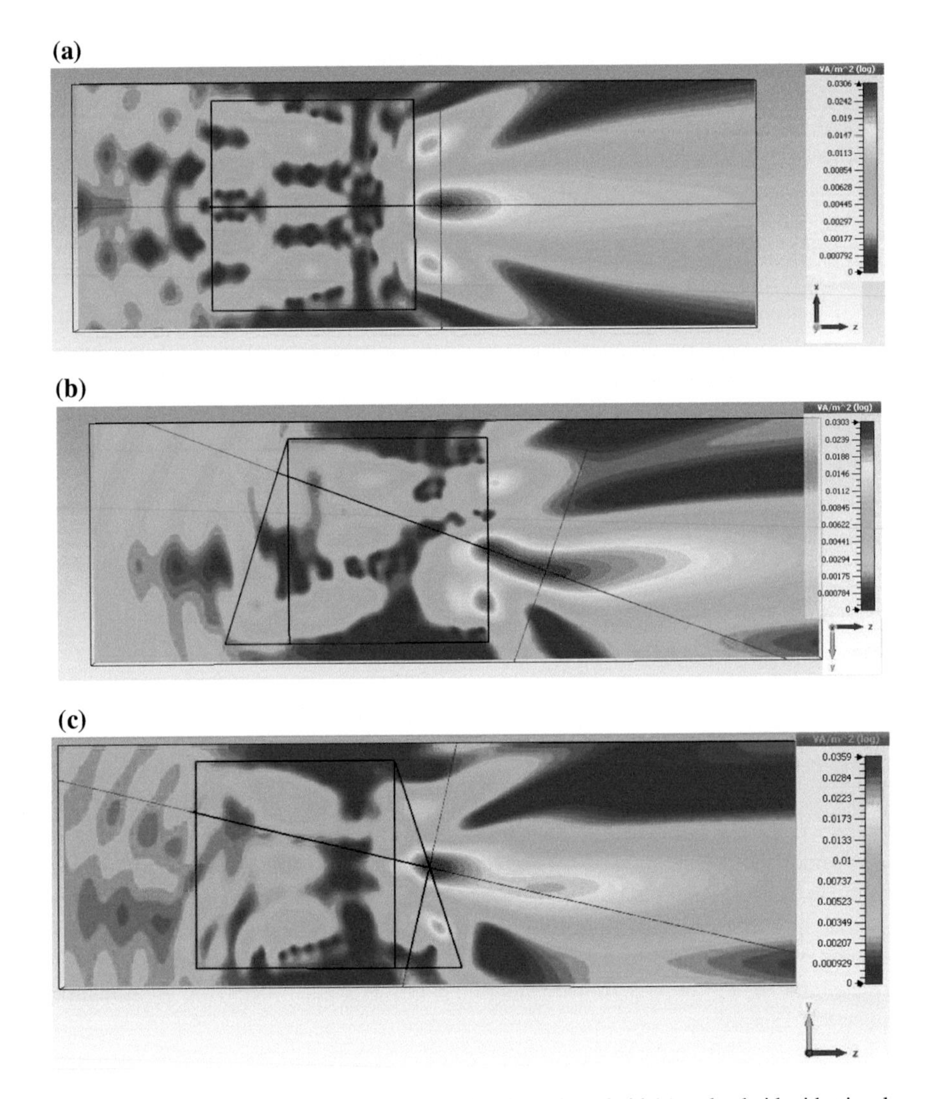

Fig. 4.7 (**a**–**c**) Simulation results of PNJ formation from the cuboid (**a**) and cuboid with triangle bar from input (**b**) and output (**c**) surface under the plain illumination., (**a**), (**b**), (**c**)

Table 4.5 Parameters of curved PNJ

Triangle position:	Intensity	Length, λ	FWHM, λ
Cuboid, normal	12.5	2.33	0.61
input	13	2.42	1.02
output	17	2.7	0.54
Cub oblique	16	1.35	0.54

Deviation angle at input angle of 18.44°. for different position of triangle bar are: 33° for cube, 20° for input and 13° for output positions of a bar. This method also allows designing a photonic jet deflector and optical trap.

Dielectric Particle of Arbitrary 3D Shape

As it was mentioned above it has been assumed that such microparticles should have high degree of spatial axial symmetry of the shape (spheres, spheroids, cylinders, disks). At the same time, to manage the whole set of PNJ parameters and optimize their characteristics additional free parameters are needed. In particular, PNJ parameters management by, for example, choosing the particle shape (cube, triangle, pyramid, etc.) has not been studied yet.

Below it was mentioned that a fundamental question is: is a hemispherical-aided shape of dielectric particle to form a photonic jet unique or the spherical shape of particle may be extended to other form? In the [37–39] as the axicon-like particle as triangle particles were considered. It could be noted that we consider the so-called mesoscale particles because of typically, this near-field focusing region (PNJ) is located at the distances not exceeding several particle diameters and is characterized by marked contribution of the radial component of optical field. In turn, this condition imposes limitations on the range of dielectric particles sizes, so it must be about a few wavelengths and *even equal to the radiation wavelength*, i.e., have a mesoscale dimensions. The specific value of the incident wavelength is not critical as long as the mesoscale conditions satisfied [37, 40].

It is important to note that the effect of internal full reflection [40] for conical particles (axicons) may destroy the photonic jet (Fig. 4.8): the conditions of non-availability of full reflection effect are $\alpha > \arccos(1/n)$ for illuminating the particle from the base and $\alpha > arctg\left(\sqrt{n^2-1}-1\right)\left(n > \sqrt{2}\right)$ (right Fig. 4.8) for the illuminating the particle from the apex. Also it could be noted that in future the criteria of different shape particle comparison (Fig. 4.9) could be developed: can we compare the particle on effective dimension calculating on its value or input cross section of particle? In this case we can more correctly compare the jet parameters for particles of different shape [40].

Thus, the principal possibility of PNJ formation in case of plane wave front interaction with the particle in the form of axisymmetric pyramid and triangular bar were shown for the first time [37–40]. The relation between the geometry (form)

Fig. 4.8 The full internal reflection effect in axicon-type particle (*left*) and dependences of critical apex angle versus refractive index contrast

and material of a dielectric particle scatterer and its resulting field distribution may be very complex. The principle possibility of generation and management of photonic jets parameters (including 3D), and (taking into account the scale effect) those of photonic nanojets, by choosing the particle shape free of axial spatial symmetry are shown. The obtained results may be used in some elements of nanophotonics of arbitrary-shape isolated particles.

Photonic Jet Formation in Mirror Regime (Flat Focusing Mirror)

Various practical problems require the creation of different types of photonic jets (photon stream) with its specific characteristics and properties. To date, all known works on the PNJ formation and their applications are based on the use of low absorbing dielectric particles in the "transmitting" regime when the photon stream is localized after incident radiation passing through the particle along the propagation direction of the incident radiation.

Below at first it has been shown that the formation of PNJ is possible in the "reflection" mode when the photon stream is localized towards the direction of incidence wavefront [37–40]. Furthermore, it is shown that by choosing the geometry of the particles may adjust the shape of photon jet and its position in the space up to the location of PNJ in direction perpendicular to incident wavefront.

The results of PNJ formation in reflection mode by rectangle particle [42] are described at [37, 40, 41]. Simulations shows that the half-width of the field intensity distribution in the region of the maximum energy concentration may be less than classical diffraction limit, i.e., it is possible to overcome the fundamental to classical linear optical diffraction limit in the "reflection" mode (note that the aim of the works [37–41] was not getting the maximum achievable performance of terahertz PNJ).

It was shown [37–41] an increase in "overflow" of energy through the edges of the particles in the direction of rotation. As a result of destructive interference, this leads to distortion of the jet shape. As an example, Fig. 4.10 shows the field intensity distribution while the dielectric particle is rotating at 10° relative to the

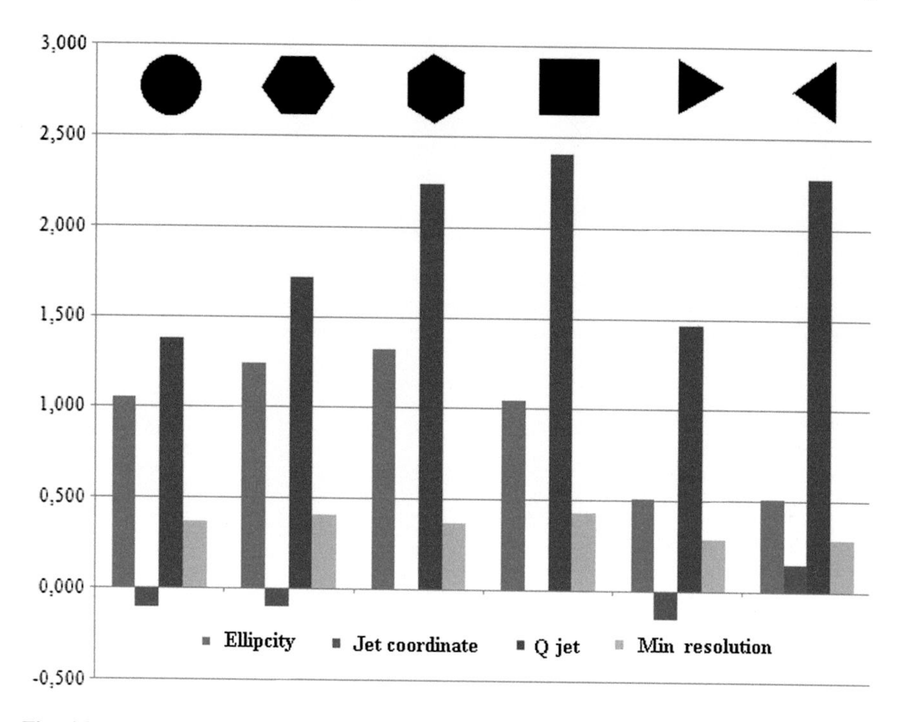

Fig. 4.9 The comparative characteristics of mesoscale particle jets: sphere, hexahedron, cubic, axicon

Fig. 4.10 PNJ Formation with increasing particle size

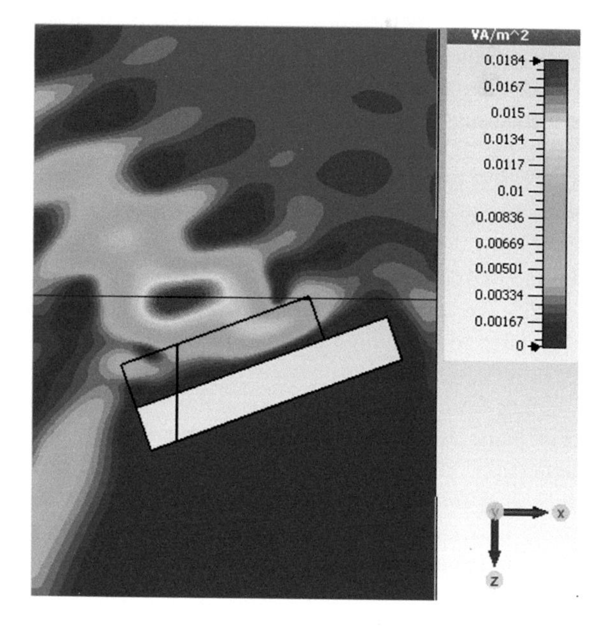

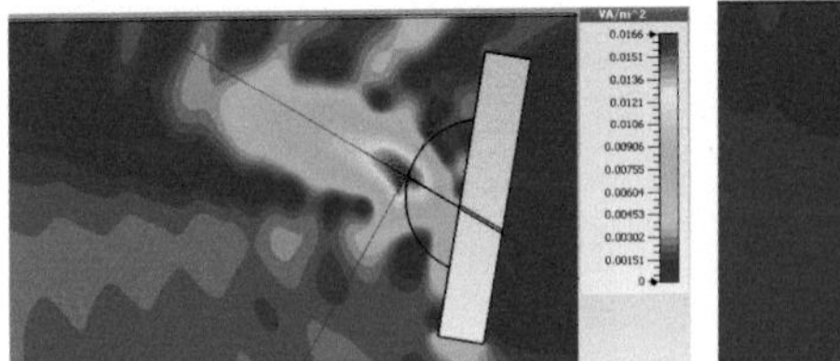

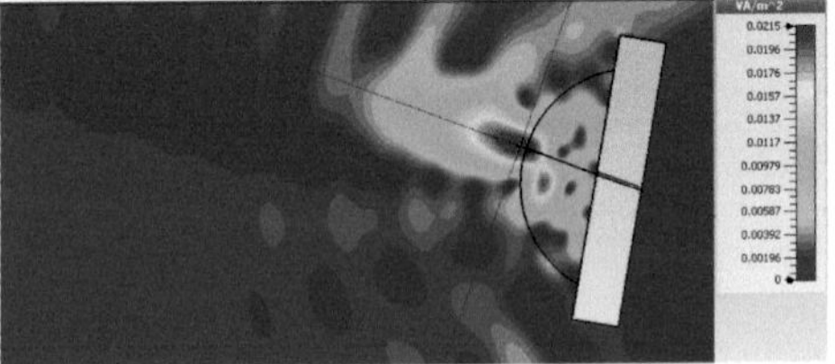

Fig. 4.11 Simulation results of the normalized power distribution on the *yz*-plane for the 3D dielectric hemisphere of radius $R = 0.53\lambda_0$ and $R = 0.8\lambda_0$ respectively (not in scale) under oblique incidence at the fundamental frequency when the whole structure is rotated at 10°

Table 4.6 Photonic jet parameters in reflection mode by semispherical particles

R, λ /angle	Δx, λ	Δy, λ	Δz, λ	Ellipticity (Δx/Δy)
R = 0.53/0°	0.44	0.44	1.04	1.0
R = 0.8/0°	0.53	0.33	0.83	1.6
R = 0.53/5°	0.44	0.47	1.05	0.94
R = 0.8/5°	0.52	0.35	0.82	1.49
R = 0.53/10°	0.42	0.46	1.05	0.91
R = 0.8/10°	0.46	0.3	0.80	1.53

direction of radiation incidence (the width of the dielectric particles has increased to a reflective edge of the substrate in the direction of rotation of the particle). As can be seen from these results, the area of high concentration of the photon stream (terajet) is not perpendicular, but substantially parallel to the flat surface of the particle. Thus, the length of the jet in this case is about (along the X axis) $1.5\lambda_0$.

The similar effects were investigated for hemispherical particles. It was shown that focusing is produced for particles with different radius, however, the focus is inside the 3D dielectric particle. By decreasing the axial cutting dimension H of particles, it is shown that the photonic jet is moved toward the hemisphere and it is close to the output surface when $H = 0.4\lambda_0$ (for $R = 0.53\lambda_0$) and $H = 0.54\lambda_0$ (for $R = 0.8\lambda_0$). As example, PNJ formations for the oblique illuminations are shown in the Fig. 4.11 and in the Table 4.6.

With the above results, it is obvious that hemispherical particles can serve as an alternative to the cuboid particles placed on reflecting substrate to form a photonic jet in reflection mode.

Thus, for the first time the possibility of photonic jet formation in the interaction of a plane wave front with a particle located on a reflecting substrate in the "reflection" mode. The principal possibility of the generation and parameters control (including three-dimensional shape and position in space) of photonic terajet (and taking into account the scale effect—and photonic nanojets) by selecting the aspect ratio and angle of rotation relative to the direction of incident wavefront are shown.

So in the area of the photonic tera-(nano-)jet applications, the undiffraction-limited information can be delivered to the dielectric layer, and the 3D dielectric particle can collect this information at the near-field of the dielectric layer. Such quasioptical transparent super-resolution imaging could apply the far-field super-resolution imaging within the area of the photonic jet. The far-field super-resolution imaging could be expected by the width and length of the photonic jet. Such photonic tera(nano)jets may need the understanding of relationship between refractive indexes and sizes of 3D dielectric particle, and effect of media and a substrate.

Photonic Jet Formation in the Scattering of Femtosecond Pulse by a Dielectric Spherical Particle

As a rule, the classical PNJ formation process corresponds to the exposure of a dielectric microsphere or cylinder to a continuous-wave radiation. At the same time, some technologies demonstrate the growing practical interest to laser systems generating ultra-short laser pulses. As a example, in [43] the authors have experimentally demonstrated the laser-induced perforation of living cell membranes using the photonic jet generated by micronsized polystyrene spheres exposed to a femtosecond pulse of Ti:sapphire laser.

In [43–45] the use of an femtosecond laser pulse, which is initially spectrally broad, for the resonance excitation of the optical field of a microsphere was proposed. It turns out that the scattering of such a pulse by the particle almost always results in the resonance excitation of the internal optical field, when the eigenfrequencies of one or several high-Q resonance modes of the particle fall within the central part of the initial radiation spectrum. In the spectral domain, the scattering of an ultra-short and hence broadband optical signal on a dielectric particle may acquire the resonance character (the particle and the ambient medium were assumed to be nonabsorbing). The PNJs, which result from the radiative loss of excited eigenmode energies through the surface of the dielectric particle, are significantly lower in intensity, but they may be subdiffractional in transverse size.

In works [46–49] based on internal optical field calculations were shown that the nonstationarity of light scattering at a transparent spherical particle manifests itself in temporal distortion of diffracted radiation and, in particular, in its temporal retarding. It has been shown that if a short-pulse laser radiation is scattered at a spherical particle, the intensity of the optical field in the PNJ area may have be pulsed in time [49]. The amplitude and the frequency of these pulsations depend on the parameters of particle morphology-dependent resonance excited by a laser pulse, as well as by the spectral detuning of these resonances from the central frequency of incident radiation.

Some Potential Applications of PNJ

The authors of [5] brief reviewed several potential applications of PNJ. Last time several interesting application were discussed in the literature, for example, optical 50 nm super-resolution [50], enhancement of Raman scattering using photonic nanojet of a single microsphere [51], microprobe based on chains of dielectric spheres [52, 53], a photonic nanojets applications for focusing electromagnetic energy into a photodiode [54], super-resolution photoacoustic microscopy [55], 2D hollow tapered waveguides [56], photonic jets with tipped waveguides [57], PNJ in double optical tweezers in an unorthodox configuration [58], PNJ array for multi-focus parallel detection of fluorescent molecules [59], a new endoscopy with microspheres for real-time white-light or fluorescent super-resolution imaging [60], nanosphere photolithography technique [61], multi-focus parallel detection of fluorescent molecules with photonic nanojets arrays [62], partial dark-field optical microscopy [63], fluorescent detection signal of on-chip immunoassays [64], enhanced Raman scattering from bulk Si and surface brilliant cresyl blue molecules [65], all-dielectric periodic terajet waveguide using an array of coupled cuboids, which may be apply to intra- and inter-chip communications and other complex systems as networks-on-chip [66], spectrometer sensor [67], THz sensors [68], extra long photonic jet [69], etc.

As an example of possible applications of photonic jet in the Fig. 4.12 the mask and field intensity distributions for 2D array of mesoscale dielectric cuboids are shown. This 2D array could be used on different fields including wavefront sensors, focal plane array, etc.

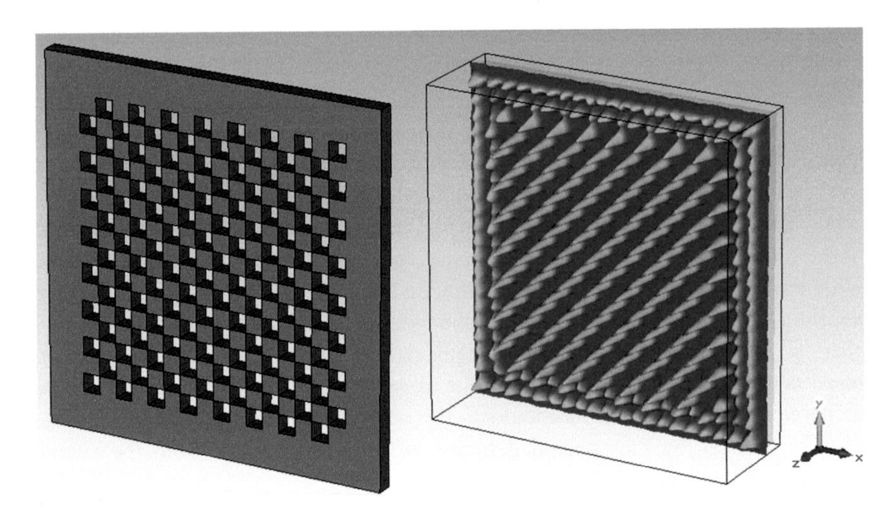

Fig. 4.12 The phases mask and field intensity distributions for 2D array of mesoscale dielectric cuboids

References

1. Born, M., & Wolf, E. (1999). *Principles of optics* (7th ed.). New York: Cambridge University Press.
2. van de Hulst, H. C. (1957). *Light scattering by small particles* (p. 470). New York: Courier Dover Publications.
3. Chen, Z., Taflove, A., & Backman, V. (2004). Photonic nanojet enhancement of backscattering of light by nanoparticles: A potential novel visible-light ultramicroscopy technique. *Optics Express, 12*(7), 1214–1220.
4. Li, X., Chen, Z., Taflove, A., & Backman, V. (2005). Optical analysis of nanoparticles via enhanced backscattering facilitated by 3-D photonic nanojets. *Optics Express, 13*, 526.
5. Heifetz, A., Kong, S.-C., Sahakian, A. V., Taflove, A., & Backman, V. (2009). Photonic nanojets. *Journal of Computational Theoretical Nanoscience, 6*, 1979.
6. Merlin, R. (2007). Radiationless electromagnetic interference: evanescent-field lenses and perfect focusing. *Science, 317*, 927–929.
7. Guo, H., Han, Y., Weng, X., Zhao, Y., Sui, G., Wang, Y., & Zhuang, S. (2013). Near-field focusing of the dielectric microsphere with wavelength scale radius. *Optics Express, 21*(2), 2434–2443.
8. Lecler, S., Takakura, Y., & Meyrueis, P. (2005). Properties of a three-dimensional photonic jet. *Optics Letters, 30*(19), 2641–2643.
9. Ku, Y., Kuang, C., Hao, X., Li, H., & Liu, X. (2013). Parameter optimization for photonic nanojet of dielectric microsphere. *Optoelectronics Letters, 9*(2), 153–156.
10. Devilez, A., Stout, B., Bonod, N., & Popov, E. (2008). Spectral analysis of three-dimensional photonic jets. *Optics Express, 16*(18), 14200–14212.
11. Mollin, R. A. (1995). *Quadrics.* Boca Raton, FL: CRC Press.
12. Jalalia, T., & Erni, D. (2014). Highly confined photonic nanojet from elliptical particles. *Journal of Modern Optics, 61*(13), 1069–1076.
13. Liu, C. (2013). Ultra-elongated photonic nanojets generated by a graded-index microellipsoid. *Progress In Electromagnetics Research Letters, 37*, 153–165.
14. Geints, Y. E., Zemlyanov, A. A., & Panina, E. K. (2011). Photonic nanojet calculations in layered radially inhomogeneous micrometersized spherical particles. *Journal of the Optical Society of America B, 28*, 1825.
15. Liu, C.-Y. (2014). Photonic nanojet shaping of dielectric non-spherical microparticles. *Physica E, 64*, 23–28.
16. Devilez, A., Bonod, N., Wenger, J., Gérard, D., Stout, B., Rigneault, H., & Popov, E. (2009). Three-dimensional subwavelength confinement of light with dielectric microspheres. *Optics Express, 17*(4), 2089–2094.
17. Kong, S.-C., Sahakian, A., Heifetz, A., Taflove, A., & Backman, V. (2008). Robust detection of deeply subwavelength pits in simulated optical data-storage disks using photonic jets. *Applied Physics Letters, 92*, 211102.
18. Zhao, L., & Ong, C. K. (2009). Direct observation of photonic jets and corresponding backscattering enhancement at microwave frequencies. *Journal of Applied Physics, 105*, 123512.
19. Ju, D., Pei, H., Jiang, Y., & Sun, X. (2013). Controllable and enhanced nanojet effects excited by surface plasmon polariton. *Applied Physics Letters, 102*, 171109.
20. Khaleque, A., & Li, Z. (2014). Tailoring the properties of photonic nanojets by changing the material and geometry of the concentrator. *Progress In Electromagnetics Research Letters, 48*, 7–13.
21. Kim, M. S. et al. (2011). Engineering photonic nanojets. *OPTICS EXPRESS, 19*(11), 10206.
22. Rahman, B. M. A., & Agrawal, A. (2013). *Finite element modeling methods for photonics* (p. 268). London: Artech House.
23. Clemens, M., & Weiland, T. (2001). Discrete electromagnetism with the finite integration technique. *Progress In Electromagnetics Research, PIER, 32*, 65–87.

24. Weiland, Thomas. (2003). Computational electromagnetics: Finite integration method and discrete electromagnetism. *Lecture Notes in Computational Science and Engineering, 28*, 183–198.
25. Pacheco-Pena, V., Beruete, M., Minin, I. V., & Minin, O. V. (2014). Terajets produced by 3D dielectric cuboids. *Applied Physics Letters, 105*, 084102.
26. Pacheco-Peña, V., Beruete, M., Minin, I. V., Minin, O. V. (2015). 3D dielectric cuboids: an alternative for high resolution terajets at THz frequencies. Proceedings EuCAP, accepted.
27. Staffaroni, M., Conway, J., Vedantam, S., Tang, J., & Yablonovitch, E. (2012). Circuit analysis in metal-optics. *Photonics Nanostructures - Fundamentals Applications, 10*, 166.
28. Minin, I. V., & Minin, O. V. (2014). Experimental verification 3D subwavelength resolution beyond the diffraction limit with zone plate in millimeter wave. *Microwave and Optical Technology Letters, 56*(10), 2436–2439.
29. Minin, O. V., & Minin, I. V. (2004). *Diffractional optics of millimeter waves*. Boston: IOP Publisher.
30. Zhao, L., & Ong, C. K. Direct observation of photonic jets and corresponding backscattering enhancement at microwave frequencies. http://arxiv.org/ftp/arxiv/papers/0903/0903.1693.pdf
31. Heifetz, A., Huang, K., Sahakian, A. V., Li, X., Taflove, A., & Backman, V. (2006). Experimental confirmation of backscattering enhancement induced by a photonic jet. *Applied Physics Letters, 89*(22), 221118.
32. Pacheco-Peña, V., Beruete, M., Minin, I. V., & Minin, O. V. (2015). Multifrequency focusing and wide angular scanning of terajets. *Optics Letters, 40*(2), 245–248.
33. Liu, Y., Wang, B., & Ding, Z. (2011). Influence of incident light polarization on photonic nanojet. *Chinese Optics Letters, 9*(7), 072901.
34. Kong, S.-C., Taflove, A., & Backman, V. (2009). Quasi one-dimensional light beam generated by a graded-index microsphere. *Optics Express, 17*, 3722–3725.
35. Bamgartl, J., Mazilu, M., & Dholakia, K. (2008). Optically mediated particle clearing using Airy wavepackets. *Nature Photonics, 2*, 675–678.
36. Polynkin, P., Koselik, N., Moloney, J. W., Siviloglou, G. A., & Christodoulides, D. N. (2009). Curved plasma channel generation using ultraintense Airy beams. *Science, 324*, 229–232.
37. Minin, I. V., Minin, O. V., & Geints, Y. (2015). Localized EM and photonic jets from non-spherical and non-symmetrical dielectric mesoscale objects: Brief review. Annalen der Physik doi:10.1002/andp.201500132 (See also Minin, I. V., Minin, O. V., & Haritoshin, N. A. (2014). Potonic terajet formation by dielectric particles of non axial symmetry of the spatial form. Vestnik SGGA).
38. Minin, I. V., Minin, O. V., & Haritoshin, N. A. (2014). Mirror photonic terajet formation. Vestnik SGGA.
39. Minin, I. V., & Minin, O. V. (2014). Photonics of isolated dielectric particles of arbitrary three-dimensional shape—a new direction in optical information technology. *Vestnik NGU Series: Information Technology, 12*(4), 69–70.
40. Minin, I., & Minin, O. (2015). Photonics of Mesoscale Nonspherical and Non Axysimmetrical Dielectric Particles and Application to Cuboid-Chain with Air-gaps Waveguide Based on Periodic Terajet-Induced Modes (*Invited*). *Proceedings of the 17th International Conference on Transparent Optical Networks, Budapest*, paper We.D6.6.
41. Minin, I. V., Minin, O. V., Pacheco-Pena, V., & Beruete, M. (2015). Localized photonic jets from flat dielectric objects in the reflection mode. *Optics Letters, 40*(10), 2329–2332.
42. Pacheco-Peña, V., Beruete, M., Navarro-Cía, M., Minin, I. V., & Minin, O. V. (2015). High resolution terajets using 3D dielectric cuboids. *The 2015 IEEE AP-S Symposium on Antennas and Propagation and URSI CNC/USNC Joint Meeting*, Vancouver, July 19–25, 2015 (accepted).
43. Terakawa, M., & Tanaka, Y. (2011). Dielectric microsphere mediated transfection using a femtosecond laser. *Optics Letters, 36*, 2877–2879.
44. Zemlyanov, A. A., & Geints, Y. E. (2004). Intensity of optical field inside a weakly absorbing spherical particle irradiated by a femtosecond laser pulse. *Optics Spectroscopy, 96*, 298–304.

45. Jipa, F., Dinescu, A., Filipescu, M., Anghel, I., Zamfirescu, M., & Dabu, R. (2014). Laser parallel nanofabrication by single femtosecond pulse near-field ablation using photoresist masks. *Optics Express, 22*(3), 3356–3361.
46. Shifrin, K. S., & Zolotov, I. G. (1994). Quasi-stationary scattering of electromagnetic pulses by spherical particles. *Applied Optics, 33*, 7798–7804.
47. Mees, L., Gouesbet, G., & Grerhan, G. (2001). Interaction between femtosecond pulses and a spherical microcavity: Internal fields. *Optics Communications, 199*, 33–38.
48. Kozlova, E. S., & Kotlyar, V. V. (2015). Simulation of the resonance focusing of picosecond and femtosecond pulses by use of a dielectric microcylinder. *Computer Optics, 39*(3), 319–3.
49. Abdurrochman, A., Lecler, S., Fontaine, J., Mermet, F., Meyrueis, P. et al. (2014). Photonic jet to improve the lateral resolution of laser etching. *Proceedings of SPIE 9135, Laser Sources and Applications II*, 913523. doi:10.1117/12.2052718
50. Wang, Z., Guo, W., Li, L., Luk'yanchuk, B., Khan, A., Liu, Z., et al. (2011). Optical virtual imaging at 50 nm lateral resolution with a white-light nanoscope. *Nature Communications, 2*, 218.
51. Dantham, V. R., Bisht, P. B., & Namboodiri, C. K. R. (2011). Enhancement of Raman scattering by two orders of magnitude using photonic nanojet of a microsphere. *Journal of Applied Physics, 109*, 103103.
52. Hutchens, T., Darafsheh, A., Fardad, A., Antoszyk, A., Ying, H., Astratov, V., & Fried, N. (2014). Detachable microsphere scalpel tips for potential use in ophthalmic surgery with the erbium:YAG laser. *Journal of Biomedical Optics, 19*, 018003.
53. Allen, K., Darafsheh, A., Abolmaali, F., Mojaverian, N., Limberopoulos, N., Lupu, A., & Astratov, V. (2014). Microsphere-chain waveguides: Focusing and transport properties. *Applied Physics Letters, 105*, 021112.
54. Hasan, M., & Simpson, J. (2013). Photonic nanojet-enhanced nanometer-scale germanium photodiode. *Applied Optics, 52*, 5420.
55. Upputuri, P., Wen, Z., Wu, Z., & Pramanik, M. (2014). Super-resolution photoacoustic microscopy using photonic nanojets: A simulation study. *Journal of Biomedical Optics, 19*(11), 116003.
56. Chen, Y., Xie, X., Li, L., Chen, G., Guo, L., & Lin, X. (2015). Improving field enhancement of 2D hollow tapered waveguides via dielectric microcylinder coupling. *Journal of Physics D: Applied Physics, 48*, 065103.
57. Takakura, Y., Halaq, H., Lecler, S., Robert, S., & Sauviac, B. (2012). Single and dual photonic jets with tipped waveguides: An integral approach. *IEEE Photonics Technology Letters, 24*, 17.
58. Neves, A. (2015). Photonic nanojets in optical tweezers. *Journal of Quantitative Spectroscopy & Radiative Transfer, 162*, 122–132.
59. Ghenuche, P., De Torres, J., Ferrand, P., & Wenger, J. (2014). Multi-focus parallel detection of fluorescent molecules at picomolar concentration with photonic nanojets arrays. *Applied Physics Letters, 105*(13), 131102.
60. Wang, F., Lai, H. S., Liu, L., Li, P., Yu, H., Liu, Z., et al. (2015). Super-resolution endoscopy for real-time wide-field imaging. *Optics Express, 23*(13), 16803.
61. Liyanage, W. P. R., Wilson, J. S., Kinzel, E. C., Durant, B. K., & Nath, M. (2015). Fabrication of CdTe nanorod arrays over large area through patterned electrodeposition for efficient solar energy conversion. *Solar Energy Materials & Solar Cells, 133*, 260–267.
62. Ghenuche, P., Torres, J., Ferrand, P., & Wenger, J. (2014). Multi-focus parallel detection of fluorescent molecules at picomolar concentration with photonic nanojets arrays. *Applied Physics Letters, 105*, 131102.
63. Kim, J. H., & Park, J. S. (2015). Partial dark-field microscopy for investigating domain structures of double-layer microsphere film. *Scientific Reports, 5*, 10157. doi:10.1038/srep10157.
64. Yang, H., & Gijs, M. A. M. (2013). Microtextured substrates and microparticles used as in situ lenses for on-chip immunofluorescence amplification. *Analytical Chemistry, 85*, 2064–2071.
65. Du, C. L., Kasim, J., You, Y. M., Shi, D. N., & Shen, Z. X. (2011). Enhancement of Raman scattering by individual dielectric microspheres. *Journal of Raman Spectroscopy, 42*, 145–148.

66. Minin, I. V., Minin, O. V., Pacheco-Peña, V., & Beruete, M. (2015). All-dielectric periodic terajet waveguide using an array of coupled cuboids. *Applied Physics Letters, 106*, 254102.
67. Minin, I. V., & Minin, O.V., Patent of Russia 2014140028.
68. Minin, I. V., & Minin, O.V., Patents of Russia 2014150306, 2014142846, 153471.
69. Minin, I. V., & Minin, O.V., Patent of Russia 201415428.
70. Valev, V. K. et al. (2012). Plasmon-enhanced sub-wavelength laser ablation: Plasmonic nanojets. *Advanced Materials*, *24*, OP29–OP35.
71. Chang, R. K., & Pan, Y.-L. (2008). Linear and non-linear spectroscopy of microparticles: Basic principles, new techniques and promising applications. *Faraday Discuss, 137*, 9–36.
72. Paganini, A., Sargheini, S., Hiptmair, R., & Hafner, C. (2015). Shape optimization of microlenses. *Optics Express, 23*(10), 13099.
73. Kiasat, Y., et al. (2013). *Proceedings of the 4th International Conference on Metamaterials, Photonic Crystals and Plasmonics* (pp. 242–243). 18–22 March, 2013, Sharjah, United Arab Emirates.
74. Hou, J., et al. (2015). Magnification and resolution of microlenses with different shapes. *Micro & Nano Letters, 10*(7), 1–4. doi:10.1049/mnl.2015.0082

Chapter 5
SPP Diffractive Lens as One of the Basic Devices for Plasmonic Information Processing

Abstract The concept of the in-plane SPP curvilinear FZP-like lens, which will have a significant impact in science and technology, is offered. The adaptation of free-space 3D binary phase-reversal conical Fresnel zone plate for operation on SPP waves demonstrates that analogues of Fourier diffractive components can be developed for in-plane SPP 3D optics.

Keywords Surface plasmon wave · Curvilinear Fresnel zone plate · Resolution power · Diffraction limit

Introduction

The wide interest to surface plasmon waves (SPP) can be attributed to their widely increasing technological importance for applications such as solar-control mirrors, subwavelength optical imaging, and sensing. Also a potential interest of surface plasmon is to produce high resolution in the proximity of the asymptote of the dispersion relation where small surface plasmon wavelengths can be achieved so SPPs wavelength can be shorter than the wavelength of radiation in surrounding media, leading to applications in sub-diffraction-limited techniques.

Surface plasmon waves can exist on the interface of a dielectric and a metallic surface and results from the interaction of radiation and surface electrons at a metal-dielectric boundary [1, 2]. To characterize the SPP, the quantities such as the wavelength, the propagation length, the depth of penetration into the dielectric and metallic media usually used [1]. For the first time, apparently, the surface waves were described by Sommerfeld for the propagation of radio waves along the Earth's surface [3].

In order to excite an SPP wave, the tangential component of the wave vector of the p-polarized incident light has to match the SPP wave vector in magnitude [4]. The analysis of dispersion equation has shown [1] that field confinement below the diffraction limit of half the wavelength in the dielectric can be achieved close to ω_{spp}.

I. Minin and O. Minin, *Diffractive Optics and Nanophotonics*,
SpringerBriefs in Physics, DOI 10.1007/978-3-319-24253-8_5

In contrast to a conventional optical or quasioptical lens, plasmonic lens is an alternative to the problems of superresolution. There are two key concepts about plasmonic lens. One is the concept of negative refractive index. The other is the transmission enhancement of evanescent waves. Good reviews of plasmonic lens are given at [5].

In Plane SPP FZP

The possibilities of SPP focusing using in-plane traditional FZP was demonstrated in [6]. The Fourier Plasmonics was proposed in [6] is a counterpart of conventional Fourier Optics for manipulating in-plane SPP waves as light in free space, which can effectively miniaturize conventional optical devices to a nanometer scale due to SPP's shorter wavelength. It is possible to reproduce all of the conventional optical devices using SPPs with a much higher resolution. Therefore, a novel SPP focusing approach using an in-plane SPP Fresnel zone plate (FZP) based on the concepts of traditional Fourier Optics was experimentally demonstrated. A conventional optical binary amplitude FZP consists of a series of concentric rings, known as Fresnel zones, that alternate in transmittance between transparent (i.e., 1) and opaque (i.e., 0) [7]. A Si-based in-plane SPP FZP, operating at the free space wavelength of 1.55 μm, was successfully fabricated on the Al/air interface using standard E-beam lithography. Two nanohole arrays were integrated on both sides of the device for launching SPP waves and visualizing their propagation [6].

For the SPP adaptation, the authors of [6] replace the wavelength λ in classical FZP [7] (see Chap. 1) with the SPP wavelength λ_{spp}, which is a function of dielectric permittivities of the metal and the surrounding dielectric (air in the experiment of [6]). The FZP_{spp} was designed to operate at the optical frequency corresponding to free-space wavelength $\lambda = 1.55$ μm and primary focal length $f =$ 80 μm. The 12 opaque zones were constructed as 5 μm-wide. The results shown SPP intensity at the focal point was about 3 times that of the input SPP wave. The field intensity at the focus of the FZP_{spp} saturates when the number of slits increases. This is due to the attenuation of surface plasmons emitted by slits several decay lengths away. The effectiveness of FZP_{spp} device was limited somewhat by the fact that even its opaque zones are partially transparent and experimentally about 30 % transmission was observed through the opaque zones. The focusing intensity is expected to be further enhanced by introducing phase reversal structure, better SPP reflectors or/and novel innovative diffractive element [7, 8].

Modulation of Surface Plasmon Polariton Using a Finite-Size Dielectric Block

In [9] an analysis of the scattering characteristics of surface plasmon polaritons using a floating dielectric block shows that the air-gap thickness between a floating dielectric block and a metal surface can be an effective dynamic variable for modulating the amplitude and phase of the transmission coefficient of the SPP. It was shown that the SPP modulation profile is not sensitive to the thickness of the dielectric block, since below an air-gap thickness, the surface bound mode plays a dominant role in transferring optical energy through the dielectric block region.

This property was applied by the authors [9] to realize a focusing surface plasmon dielectric lens with the air-gap thickness. A lens profile that makes a spherically converging wavefront was obtained from classical equations [10]. The phase and amplitude modulations by the lens at a specific position y, $\varphi(h, y)$ and $\Upsilon(h, y)$, respectively, were given by:

$$\begin{aligned} \varphi(h,y) &= \Phi(h,t(y)) - k_{spp}\,t(y), \\ \Upsilon(h,y) &= A_{spp}(l - t(y))A(h,t(y)), \end{aligned} \tag{5.1}$$

where k_{spp} is the wavenumber of the SPP eigenmode, $A_{spp}(s)$ is the modal amplitude of the SPP eigenmode that is propagated a distance s, $t(y)$ is the diffractive lens surface profile function, and l is the maximum longitudinal thickness of the lens. Both the parabolic lens and the 2π modulo Fresnel lens were designed with an air-gap thickness of 50 nm [9]. As the air-gap thickness decreases monotonically below 50 nm, the amplitude and phase modulation profiles were changed according to (5.1).

The angular spectrum representation [11] to simulate the SPP wave propagation was used (as it well known the Kirchhoff integral and the representation of the field in the form of the angular spectrum of plane waves also are widely used in the calculation of diffractive optical elements too).

A few words about the resolution using surface plasmons. It has been argued in [12, 13] that the resolution is limited when using surface plasmons on lossy metals. In [14] it was shown that the size of the image focal spot is smaller than the vacuum wavelength, but it was limited to roughly half of the surface plasmon wavelength. This result agrees with [15] where it was stated that there is a maximum wavevector k_{spp} (the wavenumber of the SPP eigenmode) given by the turning point of the surface plasmon dispersion with backbending. This limit implies that there is also a limit to the maximum intensity that can be achieved by focusing a surface plasmon. So the resolution is limited by half of a surface plasmon wavelength but the smallest wavelengths decay too rapidly to allow focusing.

Innovative Quasi-3D in-Plane Curvilinear SPP Diffractive Lens

The existence of the Kirchhoff integral analogies [1] and the representation of the field in the form of the angular spectrum of plane waves for SPP [9] can directly transfer the methods of calculating the diffractive element [7] for transformation and focusing of the SPP. It could be noted that in [14] the propagation of a monochromatic SPP waves along a planar surface was considered. In contrast with the scalar approximation, the form of a generalized vectorial Huygens-Fresnel principle for surface plasmons that includes near-field and polarization effects was developed.

The dielectric lens calculation for SPP is based on a phase modulation that occurs when SPP passes through the dielectric rectangular notch located directly on the surface of SPP propagation [9]. As it was shown the dependence of the phase of the dielectric topological defect length becomes close to linear with increasing it height. The linear relationship of the phase shift with the length of the dielectric layer allows to create the given phase distribution by changing the length and/or height of the layer.

Thus, the wavefront transformation and focusing of the SPP can be carried out using dielectric diffractive structures with varying length and a varying height located directly on the surface of the SPP propagation. So the focusing of SPP in a given region ensures the formation of the corresponding phase $\varphi(y) = \Phi[h(y), l(y)]$ at the "output" of the DOE surface. Thus, the calculation of the DOE is based on the calculation of the functions $h(y)$ and $l(y)$, ensuring the formation of a given phase $\varphi(y)$.

As described above there is the possibility of phase changes of the transmitted SPP due to changes in the step height at a fixed length. The maximum value of the length of the diffractive step was selected from the condition

$$\Delta\varphi(h) = (k_{spp}^{s} - k_{spp})h_{\max} = 2\pi, \tag{5.2}$$

where k_{spp}^{s}—SPP wave vector in the area of dielectric step. This condition provides a range of phase difference $[0, 2\pi]$ between the SPP passing through the step with the dielectric constant ε_1 and SPP propagating in the medium.

A conventional free-space flat conical FZP comprises quasi-concentric structure. From the geometric consideration for the FZP on a cone surface the boundaries (a, y) of n-th zone is determined from the quadratic equation [7]:

$$a^2 tg^2\alpha - 2a\,(htg^2\alpha - n\,\lambda\,/\,2) + h^2 tg^2\alpha - n\,\lambda\,F - (n\,\lambda\,/\,2)^2 = 0, \tag{5.3}$$

$$y = (h - a)\,tg\,\alpha. \tag{5.4}$$

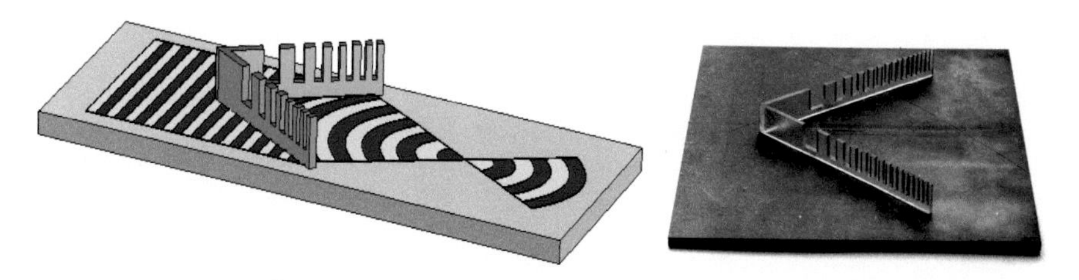

Fig. 5.1 Concept of in-plane SPP curvilinear FZP-like lens (*left*) and experimental prototype of SPP diffractive planar element on a conical surface (*right*)

Here (a, y) is function of the n, h—the height of the cone, $h = D / (2tg\alpha)$, D—is the diameter of the FZP. The design of SPP flat cone FZP (Fig. 5.1) was based on a phase modulation of the surface plasmon provided by dielectric block deposited on the interface [16].

It has been shown [8] that in contrast to the flat diffractive optics the curvilinear 3D diffractive conical optics allows for overcoming Abbe barrier with focal distance F more than $F > 2\lambda$. Moreover the longitudinal resolving power (axial resolution) of the diffractive optical element can be controlled by choosing the flexure of the diffractive optical element surface and its spatial orientation and could be less than "Abbe barrier" (see Chap. 1). So the "Abbe barrier" was completely broken by such diffractive lenses with unique 3D super resolution [8].

The successful adaptation of free-space 3D binary phase-reversal conical Fresnel Zone Plate for operation on SPP waves demonstrates [16, 17] that analogues of Fourier diffractive components can be developed for SPP 3D optics. As in free-space, the basic SPP optical components are the necessary enablers for more sophisticated future devices. It is important to mention that for longer wavelengths such as terahertz wavelengths, the air-gap dynamic range showing linear modulation characteristics may be broader. The concept of the in-plane SPP curvilinear FZP-like lens will have a significant impact in science and technology.

So we have demonstrated that analogs of diffractive and refractive 3D optics in free space can be developed to manipulate surface waves such as SPPs and focusing electromagnetic waves with diffraction limit. Demonstrating the strong electromagnetic field localization using SPP waves has been the fundamental step in the development of the opto-plasmonic subwavelength resolution optics.

References

1. Maier, S. A. (Ed.). (2007). *Plasmonics: fundamental and applications* (224 p). New York: Springer.
2. Raether, H. (1988). *Surface plasmons on smooth and rough surfaces and on gratings*. Berlin: Springer.
3. Sommerfeld, A. (1909). Surface wave. *Annalen der Physik, 28*, 665.

4. Brongersma, M. L., & Kik, P. G. (Eds.). (2007). *Surface plasmon nanophotonics* (269 p). Berlin: Springer.
5. Fu, Y., Wang, J., & Zhang, D. (2012). In K. Young Kim (Ed.), *Plasmonic lenses/chapter 8 in "Plasmonics—Principles and Applications"*. Rijeka: InTECH.
6. Feng, L., Tetz, K. A., Slutsky, B., Lomakin, V., & Fainman, Y. (2007). Fourier plasmonics: diffractive focusing of in-plane surface plasmon polariton waves. *Applied Physics Letters, 91*, 081101.
7. Minin, O. V., & Minin, I. V. (2004). *Diffractional optics of millimeter waves*. Boston: IOP Publisher.
8. Minin, I. V., & Minin, O. V. (2014). 3D diffractive lenses to overcome the 3D Abbe subwavelength diffraction limit. *Chinese Optics Letters, 12*, 060014.
9. Kim, H., Hahn, J., & Lee, B. (2008). Focusing properties of surface plasmon polariton floating dielectric lenses. *Optics Express, 16*(5), 3049–3057.
10. Born, M., & Wolf, E. (2005). *Principles of Optics*. Oxford: Pergamon.
11. Goodman, J. W. (2005). *Introduction to fourier optics* (3rd edn.). Englewood: Roberts & Company Publishers.
12. Smolyaninov, I. I., Elliott, J., Zayats, A. V., & Davis, C. C. (2005). Far-field optical microscopy with a nanometer-scale resolution based on the in-plane image magnification by surface plasmon polaritons. *Physical Review Letters, 94*, 057,401(1)–057,401(4).
13. Drezet, A., Hohenau, A., & Krenn, J. R. (2007). Comment on far-field optical microscopy with a nanometer-scale resolution based on the in-plane image magnification by surface plasmon polaritons. *Physical Review Letters, 98*, 209,703(1).
14. Teperik, T. V., Archambault, A., Marquier, F., & Greffet, J. J. (2009). Huygens-Fresnel principle for surface plasmons. *Optics Express, 17*(20), 17483–17489.
15. Archambault, A., Teperik, T. V., Marquier, F., & Greffet, J. J. "Surface plasmon Fourier optics. *Physical Review B, 79*, 195,414(1)–195,414(8).
16. Minin, I. V., & Minin, O. V. (2010). 3D diffractive focusing THz of in-plane surface plasmon polarition waves. *Journal of Electromagnetic Analysis and Applications, 2*, 116–119.
17. Minin, I. V., & Minin, O. V. (2009). 3D diffractive focusing of surface plasmon polarition waves. In *Proceedings International conference "Mathematical Physics and Nanotechnologies* (p. 96–103)". Samara, Russia, Oct.5–Nov. 6.

Chapter 6
Conclusion

Abstract The main directions of the development of diffractive optics and nanophotonics according to the authors' point of view are discussed.

Keywords on-chip optoelectronic · Mesoshape photonics · Diffractive optics · Nanophotonics · Plasmonics · Photonic jet

The progress in millimeter wave/THz/optical technology and nanophotonics industry became difficult due to a fundamental limit of radiation known as the diffraction limit.

In the current book, main attention is paid to the diffraction of millimeter wave and THz radiation on 3D subwavelength structures and also focusing of radiation below diffraction limit. The authors applied their experience in the field of diffractive optics of millimeter and THz wave to solving nanophotonics problems. Based on mathematical modelling and numerical solution of the theory of diffraction of radiation (the study of the diffraction of radiation is based on the solution Maxwell's equations), new devices containing diffractive optical elements, significantly extending, in particular, the component base of nanophotonics, were investigated and constructed.

According to the authors' point of view, the development of diffractive optics and nanophotonics are actual in the following main directions:

- Research of extraordinary effects formed in diffraction of electromagnetic waves on heterostructures containing a regular system of curvilinear steps or gaps to overcoming the diffraction limit. These researches are based on free space 3D diffractive optics development by the authors in millimeter and THz bands. It has been shown that in contrast to the flat diffractive optics the curvilinear 3D diffractive optics allows for overcoming 3D Abbe barrier with focal distance F more than $F > 2\lambda$. So an important element in the study of diffractive structures with curved areas is the creation of efficient computational methods of electromagnetic simulation.
- Of great interest is the further study of diffractive photonic crystals optics, to create nanophotonic devices that allow selective focusing of radiation with a certain wavelength and overcoming the diffraction limit. One of the advantages

I. Minin and O. Minin, *Diffractive Optics and Nanophotonics*,
SpringerBriefs in Physics, DOI 10.1007/978-3-319-24253-8_6

of using diffractive optics is the fact that multiple functions can be integrated into one element. Although, in theory several functions can be combined, the efficiency reduces with each added function. The effect of EM strong localization (EMSL) in PhC allow to desing a novel types of sensors. Photonic crystal optics allows increasing the efficiency of combined diffractive optics element.
- A photonic terajet (nanojet) formation based on diffraction effects on 3D arbitrary shape dielectric particle both in transmitting and reflecting modes are of great interest. Such each diffractive structure is created for the most effective solutions specific practical problems and is designed to best configure of nanofields. A new devices containing dielectric particle, significantly extending, in particular, the component base of nanophotonics, an on-chip optical focusing element and photonic jets produced by 3D particle of arbitrary form integrated with hollow-core fibers could be investigated and constructed.

Also, for example, some problem arises from the size incompatibility between microscale dielectric photonic devices and nanoscale silicon electronics on the chip level. The capability of photonic nanojets to generate subwavelength beam widths from dielectric particle of arbitrary 3D shape makes them good candidates to combine them together with the metallic nanostructure that produces the plasmonics in order to significantly enhance the light absorption into a semiconductor slab. A novel devices of different applications based on array of dielectric particles are also of great interest.

- The spatial distributions of localized the near-field and internal electromagnetic micro- and nanoscale intensity peaks generated at the shadow-side surfaces of arbitrary 3D shape dielectric particles which below the diffraction limit of radiation are also of great potential interest.

In additional it is interesting to study the non-linear effect in the non-linear medium around the dielectric particle, because of the intensity concentration enhances phenomenon of photonic jet.

- Transmission through a subwavelength aperture by reinforcing and localizing the incident energy at the entrance of the aperture by a nanojet or with other means of subwavelength focusing devices also of great scientific and practical interest.
- In the field of plasmonics it is interesting to study diffractive curvilinear metal-dielectric heterostructures in problems of the control of surface electromagnetic waves. In particular, it is of great interest to study heterostructures consisting of a curvilinear diffractive focusing element and grating and a uniform metal layer. Such structures can shows the formation of interference patterns of surface plasmons of different shape and their focusing in the focal area of the structure below the diffraction limit on the surface of the metallic layer.

It could be important to note that all of the practical results of this book relate to the millimeter wave and THz wave range but the methods developed can also be used for optics as a scaled model at optical frequencies, which certainly requires a large amount of additional research. Such nano-optical focusing microdevices can be integrated, for example, into existing semiconductor heterostructure platforms for next-generation on-chip optoelectronic applications.

The solution of these and other problems of the mesoshape photonics is the aim of the efforts of the authors at the present time.

It seems the works in this field compete with one another, but the priority is still with Russian scientists.

This work was partially supported by the Mendeleev scientific fond of Tomsk State University № 8.2.48.2015.

Erratum to: 3D Diffractive Lenses to Overcome the 3D Abbe Diffraction Limit

Erratum to:
I. Minin and O. Minin, *Diffractive Optics and Nanophotonics*, SpringerBriefs in Physics,
DOI 10.1007/978-3-319-24253-8_2

Erratum DOI: 10.1007/978-3-319-24253-8_7

The book was inadvertently published with an error in the title as "3D Diffractive Lenses to Overcome the 3D Abby Diffraction Limit", the correct title is "3D Diffractive Lenses to Overcome the 3D Abbe Diffraction Limit".

The original online version for this chapter can be found at DOI 10.1007/978-3-319-24253-8_2

I. Minin and O. Minin, *Diffractive Optics and Nanophotonics*,
SpringerBriefs in Physics, DOI 10.1007/978-3-319-24253-8_7

Index

I. Minin and O. Minin, *Diffractive Optics and Nanophotonics*,
SpringerBriefs in Physics, DOI 10.1007/978-3-319-24253-8